Instructor's Guide to Accompany

Delmar's Standard Textbook of Electricity 2E

By

Stephen L. Herman

and

Kerry A. Reinhackel

Delmar Publishers

an International Thomson Publishing company I(T)P®

Albany • Bonn • Boston • Cincinnati • Detroit • London • Madrid
Melbourne • Mexico City • New York • Pacific Grove • Paris • San Francisco
Singapore • Tokyo • Toronto • Washington

Notice to the Reader

Delmar Staff

Publisher: Alar Elken
Acquisitions Editor: Mark Huth
Production Editor: Megeen Mulholland
Production Coordinator: Toni Bolognino
Art and Design Coordinator: Cheri Plasse
Editorial Assistant: Dawn Daughtery

Copyright © 1999
By Delmar Publishers
a division of International Thomson Publishing

The ITP logo is a trademark under license.

Printed in the United States of America

For more information, contact:

Delmar Publishers
3 Columbia Circle, Box 15015
Albany, New York 12212-5015

International Thomson Publishing Europe
Berkshire House 168-173
High Holborn
London WC1V 7AA
England

Thomas Nelson Australia
102 Dodds Street
South Melbourne, 3205
Victoria, Australia

Nelson Canada
1120 Birchmont Road
Scarborough, Ontario
Canada, M1K 5G4

International Thomson Publishing France
Tour Maine-Montparnasse
33 Avenue du Maine
75755 Paris Cedex 15, France

International Thomson Editores
Campos Eliseos 385, Piso 7
Col Polanco
11560 Mexico D F Mexico

International Thomson Publishing GmbH
Königswinterer Strasse 418
53227 Bonn
Germany

International Thomson Publishing Asia
221 Henderson Road
#05-10 Henderson Building
Singapore 0315

International Thomson Publishing—Japan
Hirakawacho Kyowa Building, 3F
2-2-1 Hirakawacho
Chiyoda-ku, Tokyo 102
Japan

ITE Spain/Parainfo
Calle Magallanes, 25
28015-Madrid, Espana

Online Services

Delmar Online
To access a wide variety of Delmar products and services on the World Wide Web, point your browser to:
http://www.delmar.com/delmar.html
or email: info@delmar.com

thomson.com
To access International Thomson Publishing's home site for information on more than 34 publishers and 20,000 products, point your browser to:
http://www.thomson.com
or email: findit@kiosk.thomson.com

A service of I(T)P®

4 5 6 7 8 9 10 XXX 02 01

Library of Congress Cataloging-in-Publication Data 97-39869

ISBN: 0–8273–8551–X

TABLE OF CONTENTS

INTRODUCTION

This Instructor's Guide has been divided into five main sections to help you in the instruction process. The first section is an overview of instruction techniques, and information on learner types. The second section includes teaching instructions for each of the 33 units of the text. The third section includes answers and solutions to the review questions. The fourth section contains 15 tests that can be used at various intervals in the instruction process. The final section contains Tranparency Masters.

This lesson plan manual has been written with the idea that many of the instructors using Delmar's Standard Textbook of Electricity have not had an opportunity to take courses in education. With this in mind, a variety of ideas to help the instructor teach the material has been written into the section-by-section lesson plans. Additionally, this introduction will give the instructor a brief overview of the four main types of learners, and ways to enable those types of learners to be successful in the classroom. This success in the classroom will, ultimately, lead to success out in the field.

The are four basic types of learning strategies that most people use to learn new material. Some individuals may use more than one of these strategies, but most will use at least one, whether they are aware of it or not. The first type I will discuss is the visual learner.

Visual learners use what they see to help them understand what is being taught. This goes beyond just reading a text. In fact, visual learners may not benefit as much from the printed word as they do from pictures, graphs, charts, and other examples drawn **about** what is written in the text. For this type of learner there are instructions concerning visual aids throughout this entire manual. Many people are visual learners, often in combination with one or more of the other learning styles, such as audial.

An audial learner will listen carefully to everything the instructor has to say. Depending on the memory capabilities of this type of learner, he or she may also take quite a few notes. This type of learner may ask for information or explanations to be repeated, or restated in a different way. If this individual also uses the strategies of the visual learner, then explanations that are simple, clear, and are accompanied by visual aids benefit them the most. They may do just fine right where they sit. For the kinesthetic learner, though, this may not be the case.

A kinesthetic learner is a hands on type of learner. These individuals like to hold the visual aids in their hands and work things out (like experiments) with their hands. They will want to hold the magnets and pick things up with them. They will want to create a circuit and connect the wires, resistors, etc. They will also benefit, as will the visual learners, from drawing their own circuits and diagrams. They may or may not be good at comprehending the material just from reading the text, as tactical learners will be.

Tactical learners are good readers, and can comprehend most of the material that they need to learn by reading the text. Tactical learners, however, will also greatly benefit from the use of visual aids, hands on activities, and clear concise explanations.

As often as possible, the instructor should try to follow the lesson plans, bring in visual aids, make charts, and use the discussions to uncover any prior knowledge that students may have about each unit.

Individuals learn better when they believe that they already have some knowledge on a subject that they are beginning to study. With this in mind, each unit has a statement that calls for a question/discussion time that will include allowing students to realize that they probably are much more familiar with the overall subject matter than they thought they were. Think of it as wallpaper being hung on a wall. The wallpaper simply sticks better, and more readily, if there is already some paste or glue on the wall. Our minds are the same way. This prior knowledge, uncovered and made known to a student, is the glue that gives your lesson something to stick to. It sticks better, faster, and more readily when the student has been made aware that the glue is there. The slightest amount of glue is always going to hold something better than no glue at all.

Each unit is completed with a Unit Round Up section. This is where you will always need to double check for understanding. Just in case the glue didn't hold, and your lesson slid off onto the floor, you will have students explain concepts to you, work out formulas, and explain why the various things work as they do. The best way to check for understanding is to have the students give back to you, in their own words, what you have given to them. In the case of the various formulas, practice is the key. It will be useful if you can create some problems of your own for your students to solve, using the formulas that are being covered in any given unit.

This manual was written with you in mind. It is has been written in user friendly language, and is meant to be a benefit and aid to you, as a teacher. Obviously, any idea that has been given as a lesson enhancer is optional. Every learning group is different, and one group may require everything you can come up with to help them grasp the material, and other groups may barely require your presence! Use what will help you and pass over what will not.

SECTION 1
Basic Electricity and Ohm's Law

UNIT 1
ATOMIC STRUCTURE

OUTLINE

KEY TERMS

Alternating current (AC)	Electron	Nucleus
Atom	Electron orbits	Positive
Atomic number	Element	Proton
Attraction	Insulators	Repulsion
Bidirectional	Matter	Semiconductors
Centrifugal force	Molecules	Unidirectional
Conductors	Negative	Valence electrons
Direct current (DC)	Neutron	

Anticipatory Set

To prepare students for this unit, begin by stating the four objectives of the chapter. Then, ask students what they already know about any of the four things already mentioned. You may need to do a little prompting, such as asking what they remember or know about atoms, positive/negative charges, things that conduct electricity, or things that do not conduct electricity. Ask if anyone knows why you don't use electrical tools in the rain, or why you do use rubber boots protect you from shock when working on a damp floor.

1-1 Early History of Electricity

Instruct students to be prepared to define key terms in their notes as you proceed through the unit. Briefly discuss the two basic types of current. Follow up with a brief explanation of repulsion and attraction. Have some examples from list A or B (p.3 of text) to show these principles in action. Distinguish between negative and positive charges as they relate to attraction and repulsion. Ask students if they know of any other examples. Check for understanding by asking one or more students to explain the attraction/repulsion process.

1-2 Atoms

Define an atom. Discuss the three parts of the atom, and discuss the role of each part as it relates to the make-up of the 92 elements and 11 artificial elements. Check for understanding before moving on.

1-3 The Law of Charges

Begin discussion of the law of charges with a reminder of the principles of attraction/repulsion. This will help the students to immediately begin using previous learning as building blocks on which to attach new information. This is a good time to explain that everything in this text will build on principles learned, unit by unit. Discuss charges that attract and charges that repel. Use a transparency, following the outward-going and inward-going charges with some sort of pointer. Explain the two theories about why protons having the same charge do not repel each other in the nucleus. Check for understanding. Have students draw the proton-electron relationship and the proton-proton relationship in their notes.

1-4 Centrifugal Force

To begin the section on centrifugal force, ask students for examples of centrifugal force. Examples they may be familiar with include fast-spinning rides at amusement parks, or taking a curve too fast in a vehicle, or the spin cycle of a washing machine. Explain how this keeps the electrons out of the atom nucleus.

1-5 Electron Orbits

The section on electron orbits may call for a review of some basic math.

When discussing electron orbits, be sure that students understand how to determine the maximum number of electrons that can be in any given orbit. To check for understanding of the $2(N)^2$ formula, write some arbitrary possibilities on the board or overhead: $2(5)^2$, $2(3)^2$, $2(8)^2$, etc. Have students tell you how many electrons are in the orbit, and the number of the orbit, or shell: 50/5th orbit, 18/3rd orbit, 128/8th orbit, etc.

Stop here, and review the definitions of all key terms covered in the first five sections: direct current, alternating current, repulsion, attraction, atom, matter, element, electron, neutron, proton, quarks, nucleus, atomic number, opposite and like charges, and gluon. Have students explain these various terms. Review how they are related to each other.

1-6 Valence Electrons

Explain valence electrons and explain that the inner shells will have more electrons than the valence shell, as explained with $2(N)^2$. Good examples of this are the transparencies corresponding with Figures 1-15 and 1-16. Explain how atoms with fewer valence electrons make better conductors, and why atoms with a higher number of valence electrons resist conductivity.

1-7 Electron Flow

Discuss electron flow using the cue ball example. Ask students to give other examples (shuffle board, croquet, etc.). Be sure that students understand the directional flow of the energy from the impacting electron, and how that energy is split by this directional flow, lessening the energy flow in each direction.

1-8 Insulators

Review what the maximum number of valence electrons can be (8), and how this results in the creation of insulators. Have students give examples of insulators used in various professions (boots, gloves, etc.), or in household appliances (electrical cords).

1-9 Semiconductors

Explain semiconductors and give examples of how they are used (in computers, for example).

1-10 Molecules

Define molecules, and have students distinguish between molecules and atoms. See if students can explain, using what they have learned about electrons and valence shells, how two atoms might form a compound.

1-11 Methods of Producing Electricity

Go over, one by one, each of the ways to produce electricity. Discuss the benefits of using one way over another, for a particular purpose. Point out terms the student should become familiar with, such as: *piezo, seebeck, photons, photovaoltaic, photoemissive, photoconductive, solar cells,* and *cad cells.* Take some time to discuss the production of electricity from light, as it is so widely used.

1-12 Electrical Effects

Discuss how electricity can create the various things that it can be created from. Give examples as to the various uses of electricity, in these ways. For example, if you have access to a crystal radio or earphone, it would be extremely beneficial for the student to see this process in action. Perhaps you can copper plate something, or discuss the difference in buying gold plated jewelry versus solid gold jewelry. Perhaps a discussion of the way electrolysis is used would also be helpful. Having a variety of light bulbs, of different wattages, might be good way for students to explain how light is produced, and the results in heat watts and light watts.

Unit Round Up

Review all terms from the unit, including those terms covered in sections 1-6 through 1-10: valence electrons, conductor, electron flow, insulators, semiconductors, and molecules. Review basic principles covered in the unit, and check for understanding by having students explain to you how these principles work. It is also a good idea for the instructor to review the test from the Instructor's Guide to assure that all necessary material was covered well enough for students to answer all test questions.

UNIT 2
Electrical Quantities and Ohm's Law

OUTLINE

KEY TERMS

Amp
British thermal unit (BTU)
Complete path
Conventional current flow
   theory
Coulomb
Electromotive force

Electron theory
Grounding conductor
Horsepower
Impedance
Joule
Neutral conductor
Ohm (Ω)

Ohm's law
Power
Resistance
Volt
Watt

Anticipatory Set

Go over the eight objectives of this unit and ask students what they already know about any of the objectives mentioned. Be careful to correct any mistakes. For example, a student may tell you that an amp is something that he plugs his guitar into. You might even hear that an ohm is a sound you make when meditating! A little humor here relaxes the students and creates a classroom environment geared to active learning. Begin a question/answer session on the various types of measurements that students are already familiar and comfortable with. Tell them that this unit is simply an explanation of a set of measurements for electricity.

2-1 The Coulomb

Explain that the coulomb is a quantity measurement. A good visual aid is a 1 or 2 liter bottle of soda water. Show the students that it is the amount of soda water in the bottle that makes the measurement of 2 liters an accurate one. This is what "quantity" measurement means. Using that same 2 liter bottle, let it represent 1 coulomb tossed into the Pacific Ocean.

As you define Coulomb's law of electrostatic charges, write the equation on the board. Use two items to put into the Q_1 and Q_2 positions. Also, inform students what a dielectric constant is, and how dielectric breakdown can occur.

2-2 The Amp (A)

Explain that an amp is simply another measuring device: one coulomb per second. Explain the two types of measurement involved in the amp: quantity and time. Refer to the transparency corresponding with Figure 2-1 to show how an amp of current flows through a wire when one coulomb flows past a point in one second. Illustrate this by using the cue ball, or other item that will roll or slide easily. Place a coin on top of a desk and roll or slide your visual aid past the coin. Explain that the moving object represents a coulomb, the coin is the point it passes, and an amp of electricity is what flowed through the unseen wire on the desktop. So, the ampere is the measurement of the amount (quantity) of electricity flowing through a circuit. Also, refer students to Figure 2-2 in the text, making the water flow comparison.

2-3 The Electron Theory

Briefly state what the electron theory is, and draw the theory on the board showing this type of flow. Emphasize that this theory is the one that will be used throughout the text.

2-4 The Conventional Current Theory

Explain this theory carefully, making sure that students take note of the difference in flow: + to - versus - to + in the two theories. Discuss why the conventional flow theory is preferred. Use the car battery example, and check for understanding. Remember to double check for understanding on the direction that electrons **always** flow, and that the electron theory **is** the one that will be used throughout this text.

2-5 Speed of Current

Establish with students that a current is the flow of electron through a conductive substance. Refer to the example with the ping pong balls in Figure 2-6. Try to acquire a piece of 3/4" or 1" pipe, and some ping pong balls. Doing this visually in class will sink home the message of the instantaneous effect of electrical impulses. Also, remind students to try to count how long it takes a bulb to light up once a switch has been flipped. Here is a flow of electrons, the current having been activated by flipping the switch, and the speed of that flow is so fast that the bulb lights up before our hand is off the switch!

2-6 Basic Electrical Circuits

Explain closed circuits and open circuits. Emphasize the loop quality of a closed circuit. Refer back to the light switch example to illustrate how a switch is used to open a circuit when not in use, and to close a circuit when it is in use. Explain how short circuits occur, and why proper circuits and proper insulators must be used to prevent wires forming unintended short circuits. Now, add to this the explanation on grounded circuits. Have a three-prong plug or adapter to show students during this explanation. Be sure to distinguish between the grounding conductor and the neutral conductor.

2-7 The Volt (V)

Define volt for the class and distinguish it from an amp. An amp is a quantity measurement of electricity flowing through a circuit, whereas a volt is the pressure, or electromotive force, that pushes those electrons through a wire. If you refer back to the pipe and the ping pong balls, a volt is the ball that hits the last ball in line, forcing the first ball in line out of the pipe.

Explain the potential aspect of the volt and define a joule (J) for the class, letting them know that they will see this word again in the discussion of watts. Check for understanding by having students explain that voltage pushes current through a wire, but does not actually flow through the wire, as an amp does. Students should realize that voltage is potential, a force to push electricity along a current waiting to be activated. As an example, note that the wall socket has voltage, or potential, but to release that force a circuit must be activated by plugging something into that socket.

2-8 The Ohm (Ω)

Define ohm for the class. Refer to Figures 2-12 and 2-13 in the text, and use the text's example of running on the beach to explain resistance. To further illustrate resistance, use the comparison of weight lifting. The more weight you try to lift, the more your body resists the weight. Your body does this to protect you. An ohm serves a similar function in controlling the flow of electrons in an electrical circuit. Just as your muscles would short-circuit on you under too much pressure, the electrical circuit will short out with a power overload.

2-9 The Watt

Define wattage and compare it to the amp. Make the point that both of these are quantity measurements, but the watt is the power flow resulting from the electrical flow (amps) times the force applied to that flow (volts). Wattage is not a flow of current, it is a resulting amount of power.

Explain the need for energy conversion to occur before true power (watts) can exist. The electrical energy flowing through a circuit must be converted into another type of energy. Use the example of a young but extremely active child. The parent may enroll that child in several sports-related activities, to convert wild or destructive energy into focused and purposeful energy. With electricity, that conversion can be to heat energy or mechanical energy.

2-10 Other Measures of Power

Explain the definition of horsepower (hp). A drawing, even a primitive one, will be helpful. Draw a horse, with a rope hooked to a harness on the horse. Have the rope go up and through a pulley, and attach the rope to a weight. Change the poundage amount of the weight to work the formula several times. This visual aid will help students grasp the definition of horsepower. Then, show the equality of horsepower to watts.

Define British Thermal Unit (BTU), and compare it to the calorie in the metric system. Discuss what a joule is again, and that 1 joule = 1 watt. Display the transparency for Figure 2-17, and have students copy this conversion chart into the front of their notebooks for easy access, or make copies of this chart, and perhaps all conversion charts included in this text, and distribute to students.

Have students work the sample problem, then add people (a few at a time, at an average weight of 160 lbs. each), and have students figure the amount of hp needed for each new weight. Continue with these examples until the class is comfortable with this formula. Then, practice with the BTU formula and use other examples, changing the high temperature desired and/or the amount of the water to be heated.

2-11 Ohm's Law

Go over Ohm's law carefully, and have students explain what it means. Students can get confused if they don't break it down one section at a time. For clarification, go over each of the three formulas, making sure students understand that the level of resistance is equal to the number of ohms.

Next, show students how they can solve for R, I, or E, if they have the two components that make up the missing amount. For example, show students that if they know the I and R, they can find the E (E = I x R). If they need to find I, they can divide R into E. If they need to find R, they divide I into E. Practice all three of these formulas several times until students are comfortable with them.

A quick memory tool that your students can use here is the triangle:

$$\begin{array}{c} E \\ I \quad R \end{array}$$

Students can see that I= E/R, that R = E/I, and that E = R x I. It is a handy reminder of the three formulas.

Display the chart in Figure 2-20 on the overhead. Explain how the chart can be used, based on the information they have. Do several practice problems using this "Dial-A-Formula" chart, and covering the examples given in the text.

2-12 Metric Units

If your students are like most Americans, just the mention of the metric system can cause them to break out in a cold sweat. This lack of knowledge concerning the metric system can lead to learning blocks whenever the metric system is involved. As a teacher, you must help students overcome this anxiety. Explain that you are only concerned with a very basic knowledge of the metric system, and either have students copy the chart from Figure 2-24, or have a handout that they can put into their notebooks near the front for easy access. They also need to copy the chart from Figure 2-25.

Unit Round Up

Take as much time as is necessary to review this unit. It is a vitally important unit. Students need to be able to work with the various concepts and formulas comfortably and use them interchangeably. Go through the summary together, and assign the review questions to double check for complete understanding. Check these assignments before moving to the next unit to see if any reteaching needs to be done. Review all key terms from the unit.

UNIT 3
STATIC ELECTRICITY

OUTLINE

3-1 Static Electricity
3-2 Charging an Object
3-3 The Electroscope
3-4 Static Electricity in Nature
3-5 Nuisance Static Charges
3-6 Useful Static Charges

KEY TERMS

Electroscope	Lightning bolt	Selenium
Electrostatic charge	Lightning rod	Static
Lightning	Nuisance static charges	Thundercloud
Lightning arrestor	Precipitators	Useful static charges

Anticipatory Set

Unit 3 is a very basic, simple unit. Go over the objectives and ask students to share experiences they have had involving static electricity. Have some visual aids on hand to help illustrate the principles of electrostatic charges. These aids might include a rubber rod, a piece of wool, a plastic comb, some tissues, an inflated balloon, and some pictures you copied on a copy machine. When you make these copies, make them progressively darker so that you can show the process in even more vivid detail.

3-1 Static Electricity

As you discuss the various uses of static electricity, and include the explanation of precipitators, ask students to name a household item that may use this process (an air conditioning unit). Be sure that students understand what static means, and that it is a charge, not a current. It is similar to voltage in that it is a charge waiting to be released, like a volt is potential force waiting for a current to push. Also, point out the electron theory in action here.

Be sure that students realize that insulators hold electrons, and that is why electrostatic charges build on them. Make sure students understand that a lack of electrons means a majority of protons and, thus, a positive charge. The inverse of this also needs to be explained.

3-2 Charging an Object

Explain that this is merely an extension of the principles discussed in the first section. Here, we use these principles to electrostatically charge items by choice.

Have students experiment with visual aids, building up static charges. The comb can be run through someone's dry hair, and then used to pick up pieces of the tissue. The balloon can be rubbed on someone's dry hair (or on the wool), and then suspended on a wall.

3-3 The Electroscope

Describe an electroscope and explain its function. If it is possible, to have an electroscope in the classroom. Students could use the various things they used earlier, and determine what type of charge the various items took on.

3-4 Static Electricity in Nature

Define lightning as nature's static electricity, and discuss why the flow of electricity sometimes moves from the clouds to the ground, and why it goes the other way at other times. Again, the electron theory explains this upward or downward movement.

Explain what lightning rods are, and make the distinction between them and lightning arrestors. Refer to Figures 3-9, 3-10, and 3-11 in the text to help illustrate those explanations.

3-5 Nuisance Static Charges

Basically, just briefly touch on this, referring back to the things you did in Section 3-2. These principles have already been discussed. However, ask your students to explain why the various types of products that can be put in the dryer with wet clothes (Bounce, Downy Dryer Sheets, Cling, etc.), prevent the static from building up in the clothes. Ask them why an aerosol spray product made for this purpose can eliminate static cling when sprayed directly on the clothing. Ask why some clothing items cling together, while others do not (e.g., rayon socks do, but denim jeans do not; pantyhose do, but heavy towels do not).

3-6 Useful Static Charges

Discuss the various useful ways in which static electricity is used. When explaining how the copy machine works, display your series of progressively darker copies. Have students explain how the copies became progressively darker (beyond the obvious answer of pushing the button to make them darker).

Unit Round Up

Summarize the unit, review all key terms, and double check for understanding. If necessary, go back through any previous explanations, using the visual aids.

UNIT 4
MAGNETISM

OUTLINE

KEY TERMS

Ampere-turns
Demagnetized
Electromagnets
Electron spin patterns
Flux
Flux density

Left-hand rule
Lines of flux
Lodestones
Magnetic domains
Magnetic molecules
Magnetomotive force (mmf)

Permeability
Permanent magnets
Reluctance
Residual magnetism
Saturation

Anticipatory Set

This is another unit in which students will greatly benefit from several visual aids. Gather magnets of various sizes, some cardboard (even several sheets of paper stacked together will work), some paper clips, and perhaps a model, or actual, electromagnet. Use the transparencies for Figures 4-2, 4-5, 4-6, 4-7, 4-10, 4-11, 4-14, 4-15, and 4-17. State the unit objectives.

4-1 The Earth Is a Magnet

Give the explanation of true geographic north versus magnetic north. Have students draw Figure 4-2 into their notes, labeling the four poles, the polar axis, and magnetic flux lines. After they draw Figure 4-2 into their notes, explain why a compass points toward true geographic north instead of to magnetic north. Have students look at their drawing as you explain this.

4-2 Permanent Magnets

Define permanent magnets, and have some to show your students (even if they came off your refrigerator and have a phone number for ordering pizza on them, or a pig, or a cow, etc.). Be sure students write the definition of a permanent magnet into their notes, as well as the basic law of magnetism given in this section.

4-3 The Electron Theory of Magnetism

Explain electron spin patterns and how they affect, or account for, the magnetic qualities or lack of magnetic qualities in an object. Refresh students' memories by reminding them that the valence shell and, therefore, the valence electrons are at the outermost contact points of an atom. The valence shell is the outermost shell, or orbit, of an atom.

Explain magnetic domains or magnetic molecules, and how these are formed by iron, nickel, and cobalt sharing electrons in a way that does not cancel out their magnetic fields. Have students refer to Figures 4-5 through 4-7, and use these transparencies.

4-4 Magnetic Materials

Discuss the three classifications of magnetic materials. Explain what an alloy is, and why it makes better permanent magnets. Perhaps students can figure it out for themselves, just in knowing what the alloy Alnico 5 is made of. Refer them to the Ferromagnetic materials list, and ask them to explain why Alnico 5 makes such a good permanent magnet.

Discuss ceramic magnets and, if possible, show some examples to the class.

4-5 Magnetic Lines of Force

Explain what flux is, and have students make note of its symbol in their notes. If you have the magnets and iron fillings, this is the time to use them to display flux lines in action. If you do not have these items, use transparencies of Figures 4-10 and 4-11 to illustrate this process. Make sure students understand that magnetic lines of flux never cross each other, but repel each other instead. Also, have students make note of another basic law of magnetism: like poles repel and unlike poles attract.

4-6 Electromagnetics

Explain how a magnetic field is formed, and how electrical current is necessary to produce the magnetic field. Remind students that permanent magnets maintain their magnetism without electrical current, whereas electromagnets usually do not.

Illustrate the lines of flux using the transparency for Figure 4-15. Also, draw a coiled wire on the board and describe why this coiled effect produces a stronger magnetic field.

Describe the different types of core material, and have students explain why they are called air core or iron core. Additionally, have students explain why a magnetic material added to the core creates a stronger magnetic field. Define permeability, and be sure students understand how this affects the flow of electrons through the core material. Define reluctance, and discuss how it compares to permeability. Have students give examples of materials that they think would have low permeability and high reluctance. These examples could include wood, rubber, cloth, etc.

Explain what saturation is and how it occurs. Then explain residual magnetism and how a magnetized material can be said to have coercive force and retentivity.

4-7 Magnetic Measurement

Go very slowly and carefully through each of the three systems of measurement. Be sure that students write formulas and equivalents into their notes. Check for understanding.

4-8 Magnetic Polarity

Hold a pipe or rod that you have wrapped a coil of wire around. Using your left hand, demonstrate how to point, with your thumb, to the magnetic north pole. If a pipe or rod is not available, use a pencil, and wrap a string loosely around it.

4-9 Demagnetizing

Explain how the molecules must be returned to a condition of disarray in order to demagnetize the (electrically) magnetized material. Explain how this is done, referring back to coercive force, and to transparencies 4-5 and 4-6, for clarification.

4-10 Magnetic Devices

Together with the class, list some of the more common devices that operate on magnetism. Explain how the speaker in a sound system works, and ask for personal experiences that students may have had with various types of electromagnets.

Unit Round Up

Go over the summary and the review with the class. Check for understanding, being careful to ask students to explain formulas and conversions. This is always the best way to check for clear understanding.

UNIT 5
RESISTORS

OUTLINE

KEY TERMS

Carbon film resistor

Color code

Composition carbon resistor

Fixed resistors

Metal film resistors

Metal glaze resistor

Multiturn variable resistors

Pot

Potentiometer

Rheostat

Short circuit

Tolerance

Variable resistor

Voltage divider

Wire wound resistor

Anticipatory Set

This is a short unit, but could be confusing if not taught slowly and carefully. Handouts and visual aids will greatly improve student understanding of and ability to work comfortably with the information in this unit. A collection of various sizes and shapes of resistors should be readily available for student observation. Additionally, a color chart, such as is given in Figure 5-9, should be made available for each student. These should be laminated, if possible. Students could be assessed a charge for the cost of doing this, if you need to do so. The discussion of resistors will be clearer if students have their own charts to refer to, as well as actual resistors to look at.

State the objectives of the unit and check what students already know, or think they know, about resistors, and where they think resistors might be used. Allow students to examine the resistors that you brought to class. Hand out the charts, with a brief explanation of what they are, and suggest that they can be referred to throughout this section.

5-1 Uses of Resistors

Explain the two functions of resistors, and why the resistors are used these two ways. Use transparencies that correspond with Figures 5-1 and 5-2 in the text. Within this explanation, discuss how a short circuit is created (as a result of no resistor being present), and how voltage is divided using resistors.

5-2 Fixed Resistors

Define fixed resistors, what they are made of, and why they are so popular. Compare carbon resistors with metal film resistors, especially regarding change of value. Be sure students understand that the value spoken of with resistors has to do with the level at which a resistor will resist an electrical flow. The lower the tolerance level, the higher the resistance. With carbon resistors this tolerance can change, but with metal film resistors it does not.

Describe carbon film resistors and how, just as the name would indicate, these resistors include the best of both of the other two resistors, being relatively inexpensive and having a higher tolerance rating.

Next, expand this discussion to include metal glaze resistors and wire wound resistors. Explain how the resistance of a wire wound resistor is figured. Be sure students note that this type of resistor can function at higher temperatures than other types of resistors. Explain the need for a vertical mounting of these resistors to allow air to circulate. For now, just mention inductance as a variation in current that could wreak havoc on a circuit, and leave a more detailed definition for later units.

5-3 Color Code

Use the charts in this unit (color coded as in Figure 5-9). Go over how the value of a resistor can be determined, and have students refer to their charts. Explain how the color code works, and go over the example given on page 99 of the text. This will familiarize students with the formula for determining a resistor's value and tolerance.

Explain the differences between these basic processes for figuring value and tolerance, and the processes that must be followed if the multiplier band is gold or silver. Work a few problems using both of the formulas and the chart to familiarize students with both procedures.

5-4 Standard Resistance Values of Fixed Resistors

Go over the standard values chart on page 104 of the text, and be sure that students understand why the 2% to 5% section only has 24 values listed, while the 1% section has 96 values. Also inform students that any of these may be multiplied by 10 to create other values.

5-5 Power Ratings

Explain power ratings and what can happen if a resistor's power rating is exceeded. Go over the three formulas that can be used to figure a resistor's power rating, and do some practice problems for each formula on the board. Emphasize the danger of exceeding a resistor's power rating.

5-6 Variable Resistors

Explain variable resistors, making the obvious distinction between them and fixed resistors. Be sure students understand that whatever the maximum resistance value is, the variable resistor can be set between 0 and that maximum number. Have students explain why this type of resistor might be advantageous over a fixed resistor.

Discuss the three-terminal wiper arm type of variable resistor. Discuss multiturn variable resistors and compare them with resistors that are limited to a 3/4 turn. Also have students make note of the fact that the wire wound type of resistor allows for a higher power rating. As always, any examples that you can make available to your students will be invaluable as teaching aids to illustrate the points that you are making.

Go over the terminology and have students put these terms into their notes. Define each one and check for understanding.

5-7 Schematic Symbols

Use transparencies for this. After going over them, have students practice drawing them into their notes.

Unit Round Up

Go over the summary and review together, and check for understanding. Make copies of the resistors chart on pages 96-97 of the text, and have students complete this chart in class as your check for understanding. Go over any areas that students do not appear to understand clearly.

SECTION 2
Basic Electrical Circuits

UNIT 6
SERIES CIRCUITS

OUTLINE

KEY TERMS

Chassis ground
Circuit breakers
Earth ground
Fuses
Ground point

General voltage divider formula
Ohm's law
Resistance adds
Series circuit
Series resistance

Voltage divider
Voltage drop
Voltage polarity

Anticipatory Set

Prepare students for this lesson by first reviewing the previously learned formulas that will be used in this unit. These include: I = E/R, E = I x R, and R = E/I. Review the role of resistors and, if needed, review the practice chart from page 111 and double check for understanding.

Use the transparencies that correspond with Figures 6-1 through 6-25 on the overhead. Go over each of these carefully, calling on a different student for each blank that is to be filled in.

This is a pivotal point in the learning process. An accounting of all previously learned material is necessary in this unit. A complete working understanding of the material covered up through the end of this unit is crucial to moving forward.

State the objectives and ask for prior knowledge of, or experience with, circuits, circuit boards, etc.

6-1 Series Circuits

Define a series circuit and remind students that this text bases explanations on the electron theory (have a student explain this theory). Explain how fuses and circuit breakers are used in a series circuit. Ask if anyone has experience with blowing out fuses or tripping circuit breakers (perhaps by plugging in one too many things). Explain why it is not a good idea to bypass these devices with foil, paper clips, pennies, etc.

6-2 Voltage Drops in a Series Circuit

Define voltage drop, and use transparencies 6-4 through 6-6 to help explain what this means. Go over each of these transparencies slowly, checking for understanding as you go.

6-3 Resistance in a Series Circuit

Explain how the total resistance is figured, and remind students that this is in a series circuit. Show this to the class, using transparencies 6-4 and 6-5.

6-4 Calculating Series Circuit Values

Explain the three rules for calculating series circuit values, and refer to transparency 6-5. To check for understanding, have students determine what amperage the current is flowing at in transparency 6-5. If needed, remind them that $I = E/R$. Also, have them figure the R_T for the same circuit. Use transparencies 6-6 and 6-7 to further illustrate this process. Be sure students understand that the $E_T = E_1 + E_2 + E_3$.

6-5 Solving Circuits

Explain briefly that with any series circuit, if at least two factors are given, the missing factor can be found using the rules for series circuits and Ohm's law. Explain each formula and have students fill in the missing numbers for Figures 6-8 through 6-11. This is an excellent way to test for understanding.

Be sure to explain why you used $P = E \times I$ with one set of factors, and $P = E_2^2/R_2$ with another. Explain that the answer will be the same, but one formula is easier to use with decimals than the other. Also, as you progress you may need to do a brief algebra review to figure missing factors for circuits in Figures 6-12 through 6-20. For example, Figure 6-17 provides P_T, and students will need to do some very basic math to solve for P_1. Show the students that once they have found P_1 and E_1, they can find I, which is a constant throughout a series circuit. To find I, they will need to remember that $P = E \times I$. So, write it out for your students on the board or overhead:

$P = E \times I$

$.205 = 6.4 \times I$

$.205/6.4 = 6.4/6.4 \times I$

$.032 = I$

You may need to remind students that in an equation, whatever you do to one side of the equation, you must also do to the other side. The key here is that I is multiplied by 1, so the answer is simply I.

Have students fill in .032 for I_1, I_2, I_3, I_4, and I_T. They now have enough information, using Ohm's law and the three rules for figuring series circuit values, to fill in all the blanks. Prepare some circuits of your own, with just enough information to get them started, and have them do several practice circuit problems before advancing to voltage dividers.

6-6 Voltage Dividers

Explain what a voltage divider is and how it works. Refer to the transparency for Figure 6-22. Before turning to page 134 of the text, give students the information printed on that page for the 120V, 0.015_{mA} circuit. Have students draw this circuit, and fill in E_1, E_2, E_3, E_T; I_1, I_2, I_3, I_T; R_1, R_2, R_3, R_T; and P_1, P_2, P_3, P_T. Using the formulas they have learned so far, they should be able to draw this circuit and fill in all the blanks. This is an excellent check for understanding, and will let you know if anything needs to be retaught. Also, go over the metric conversion chart again.

6-7 The General Voltage Divider Formula

Explain and demonstrate the formula. Explain that it is closely related to the E-R-I triangle. Have students do some practice problems. As with all formulas, have students copy this into the front of their notebooks. Also, cover the proper procedure for rounding off numbers. Explain why it is done and how to know whether to carry the answer beyond three place values to the right of the decimal.

6-8 Voltage Polarity

Explain this principle, and use the transparency for Figure 6-22 to help illustrate what you are saying. Basic language such as, "the point closest to the negative terminal in any given set of points is always going to be more negative than the point farther away" seems like common sense, but use the transparency to check for understanding anyway.

6-9 Using Ground as a Reference

Display the symbols for earth ground and chassis ground on the board or overhead, and explain the difference. Use the automobile chassis as an example of a chassis ground, as is done in the text. For an example of an earth ground, refer back to Unit 3 and the discussion of lightning rods.

Discuss above ground voltage versus below ground voltage, with ground serving a type of dividing line or common point of separation. Remind students that an above ground voltage is positive in respect to ground, and that a below ground circuit is negative in respect to ground.

Unit Round Up

Do the summary and review in the text orally to assess the student's level of understanding of this unit. Also, provide copies of the circuits chart on page 140 of the text for students to complete.

UNIT 7
PARALLEL CIRCUITS

OUTLINE
7-1 Parallel Circuit Values
7-2 Parallel Resistance Formulas

KEY TERMS

Circuit branch
Current adds
Current dividers

Load
Parallel circuits
Product over sum method

Reciprocal formula

Anticipatory Set

Review resistors, Ohm's Law, and the rules of series circuits. Explain to students that Units 7 and 8 will call for them to be familiar with all of the previously learned formulas and their applications. The new formulas learned in this unit will be used as well in Unit 8.

Instruct students to be ready to write notes concerning parallel circuits into their notebooks. Go over the unit objectives and make students aware of the key terms they will need to learn in this unit.

Prepare some blank parallel circuit diagrams, making them progressively more complex, for students to practice on. Fill in just enough information for the student to start with, and then have them use the formulas to fill in the blanks. These circuit diagrams make excellent warm-up exercises for Unit 7.

7-1 Parallel Circuit Values

Total Current

Define what a parallel circuit is, and either draw Figure 7-1 on the board, or use the corresponding transparency. Students need to *see* this definition, as well as hear it. Using this same transparency, explain current adds. This is much easier to understand visually than it is verbally. Have students explain this concept back to you, using the transparency.

Voltage Drop

Use transparency 7-2, or draw Figure 7-2 on the board, and point to the voltage across each branch of the circuit. Have students copy the rule for voltage drop into their notes. Compare this to a series circuit, where the amperage (I) is a constant throughout the circuit. So, if the E_T is known, then E_1, E_2, E_3...E_n can be filled in immediately in a parallel circuit (as I can be, in a series circuit), because they will be the same as E_T.

Total Resistance

This may be a little difficult for students to grasp at first. Explain the rule, and use the water pump and holding tank example. Another example is a garden hose that has holes in it. Without the holes, the overall resistance to the flow of water through the nozzle is much higher than it will be when the hose has holes in it. Every time a new hole is made in the hose, the overall resistance drops again.

Ask students to visualize bending the hose to cut off the water flow. Now the resistance is high. When they open it back up and give the water another outlet (like a parallel circuit), then the overall resistance drops.

So, in a parallel circuit, since the voltage (the pusher of the current) stays constant, with less resistance but the same push, the rate of the flow of that current will increase. This increase is what leads to load. Be sure students can explain this complete process to you before moving on to Section 7-2.

7-2 Parallel Resistance Formulas

Resistors of Equal Value

Explain how the formula works and emphasize that it can only be used when all resistors are of equal value.

Product Over Sum

Go over this slowly, and work a few on the board for clarification. Be sure that students understand the substitution process when figuring the overall resistance. Have students do some of these on their own, until they are comfortable with the formula.

Reciprocal Formula

Explain that this formula is ideal for situations when a missing resistance value prevents the individual from using the product over sum method. Go over this formula carefully, step by step. Have students use the formula to figure out the missing values for Figure 7-15. Remind students that once they have found any E value in the circuit, E is a constant in a parallel circuit. Also remind students of Ohm's law, and the R-E-I triangle. Another visual aid they can use is the P-I-E triangle. Have students draw them on their worksheets for easy reference:

Current Dividers

Explain this process and have students refer to Figure 7-23 in the text. Take students step by step through the current divider formula. Show them the substitution process (I_T x R_T for E_T). Before allowing students to look at Figure 7-25 and the solution, draw this circuit on the board and see if students grasp the substitution process, as well as the concept that I_x represents any location, for current, on the circuit. This applies to R_x as well.

Unit Round Up

Go over the summary in class. Spend as much time as is necessary doing the review, working out the sample problems, and having students fill in the Practice Problem chart. Remember, practice is the best way for students to learn and become comfortable with these processes.

UNIT 8
COMBINATION CIRCUITS

OUTLINE

KEY TERMS

Combination circuit

Kirchhoff's current law

Kirchhoff's law

Kirchhoff's voltage law

Node

Norton's theorem

Parallel block

Redraw

Reduce

Simple parallel circuit

Superposition theorem

Thevenin's theorem

Trace the current path

Anticipatory Set

This unit could prove to be a very difficult one for some students. It will require a time frame sufficient to include plenty of practice time and algebra review time. As you progress through the unit and the combination circuits become more complex, provide some blank practice circuits for students to work on.

As you move into Kirchhoff's law, Norton's theorem, Superposition theorem, and on into Thevenin's theorem, be sure to take as much time as the class needs to grasp the concepts and become comfortable working the problems. This is extremely important before moving on. A good technique to use with this unit is to allow students to work out problems on the board, so that you can see where the confusion is, should there be any.

8-1 Combination Circuits

Have students trace, with their fingers, the path of the current as it leaves point A and travels to point B (Figure 8-1). Identify what a node is, as part of a combination circuit, and define combination circuit.

8-2 Solving Combination Circuits

Review the rules for solving series circuits and the rules for solving parallel circuits. Have students make flash cards of these rules on 3 x 5 index cards to keep nearby while solving combination circuits.

8-3 Simplifying the Circuit

This is a crucial concept that must be fully understood in order for Unit 8 to be completed successfully. Go over the reciprocal formula again, and carefully walk students through the first two simpler examples. Then let them wrestle with Figure 8-9 (or a similar practice example). Provide copies of Figure 8-9 to students as work sheets. After they have had a reasonable amount of time to solve this circuit, have students explain each step (unless no one solved it, then you will need to solve it with them, and give them another one to try). Make sure that students understand both how to simplify the circuit into a series circuit, **and** how to break the circuit back out into a combination circuit, solving for parallel values along the way.

Kirchhoff's Law

Distinguish between voltage and current applications, and demonstrate both of these in action. Have students trace the circuits in the examples given, reminding them that the assumption is that current flow moves from negative to positive. Explain the negative/positive relationship of current entering and leaving a resistor. Practice solving some circuits using Kirchhoff's law, and emphasize why you would use these methods, rather than Ohm's law, in these situations. Let students practice using the rules of Kirchhoff's law as long as they need to in order to feel comfortable with the process.

Thevenin's Theorem

Explain how this works and, if possible, have a black box to use, as a visual aid. Talk students through using Thevenin's theorem by doing several on the board. Have students take turns coming to the board and explaining the process. Remind students that they may still use series circuit and parallel circuit rules once a circuit has been simplified. Be sure to go over the Thevenin equivalent circuit.

Norton's Theorem

Explain how the theorem works, and when it is most helpful in solving combination circuits. Have students practice solving several problems using Norton's Theorem. Distinguish it from Kirchhoff's law and Thevenin's theorem, discussing when one is more beneficial to use than the other. Compare the Norton equivalent circuit to the Thevenin equivalent circuit.

Current Sources

Explain the two different ways (voltage or current) that a power source can be identified. Students should be completely familiar, by now, with the V and A indicators for volts and amps.

Determining the Norton Equivalent Circuit

Explain the process, describing the role of the short circuit as seen in Figure 8-36. Make sure students understand how to do this, why this process is useful, and where it will be used.

The Superposition Theorem

Explain that students will use in this next theorem the three algebraic processes they have already learned. Remind students that in this theorem, as in all the others, the key to simplifying the solving process is circuit reduction. Talk students through this process using circuit diagrams on the board or overhead. You may need to go through this several times before it is clear.

Unit Round Up

Do the summary together, in class. Assign the review for homework, to be sure that students can use what they have learned in Unit 8 away from the classroom. Have students solve the practice circuit problems that follow the review questions. Go over all of these in class, having students explain what they did to achieve the answers they came up with. Review all new terms covered in Unit 8.

SECTION 3
Meters and Wire Sizes

UNIT 9
MEASURING INSTRUMENTS

OUTLINE

KEY TERMS

Ammeter	Current transformer	Multirange voltmeters
Ammeter shunt	D'Arsonval movement	Ohmmeter
Analog meters	Galvanometers	Oscilloscope
Ayrton shunt	Moving coil meter	Voltmeter
Bridge circuit	Moving iron meter	Wattmeter
Clamp-on ammeter	Multirange ammeters	Wheatstone bridge

Anticipatory Set

Have as many of these measuring devices as possible in your classroom. The combination of hearing the explanation of how they each work while watching them work is hard to beat as a teaching strategy.

Encourage students to take a few notes about each device, especially the ones they are not familiar with. Those that students are familiar with could be discussed first, especially if Unit 8 was a little difficult. Having devices students are familiar with will allow them to feel that they already know something about the material to be covered in Unit 9. This type of atmosphere always lends itself to better learning.

Begin by going down the list of measuring devices named under Key Terms, and point to the ones you have available. If it is possible to do so, allow students to work these devices in class. As always, go over the unit objectives, and alert students to the key terms they will be responsible for in this unit.

9-1 Analog Meters

Principle of Operation

Point out how they are distinguishable (pointer and scale). Discuss the different types, and demonstrate with any you have in class. Be sure students understand how the current creates the magnetic field that causes these meters to operate. A brief review of Unit 4 might be helpful before moving forward.

Moving Iron Meters

Discuss the three types and have students view Figures 9-4 through 9-6, as well as any you have in the room. Discuss how the magnetic field is created. Spend a little time explaining AC/DC currents, rectifiers, and why only direct current works with these meters.

9-2 The Voltmeter

Explain what a voltmeter is, and demonstrate it, if you have one. Be sure students understand that the leads are connected *across* the power source. Also, have students make note that it can be used to measure DC or AC, and explain the meter circuit (meter plus resistor).

Calculating the Resistor Value

Explain what the *full scale values* of the meter are. Explain how the resistor value is figured, using Ohm's law: I x R = E; E/I = R; E/R =I.

9-3 Multirange Voltmeters

Discuss these voltmeters, demonstrate one if possible, and discuss increase and decrease of resistance as it applies to the volt level chosen.

Calculating the Resistor Values

Demonstrate on the board how students can determine the resistor values by using Ohm's law. Point out that the total volt value per scale is what will cause the resistor values to change. If possible, move the switch on a voltmeter, and have students solve for the resistor values.

9-4 Reading a Meter

Go over the various things that can be measured on a multimeter. Either demonstrate these things on a real multimeter, or have students refer to Figure 9-13 in the text.

Reading a Voltmeter

Point out the various scales on the multimeter, and explain the process of multiplying or dividing by 10 to achieve an accurate reading. Take students through the three steps for reading a meter, and be sure they understand how to do each one. This is where an actual multimeter is invaluable as a visual aid. Adjust the setting, and then have students practice recording the values, using 10 to multiply or divide to achieve an accurate reading.

9-5 The Ammeter

Explain what an ammeter is used for, and either have one in the room or refer students to Figure 9-17 in the text. Emphasize the necessity for the ammeter to be connected in series with the load, and what the result would be if this isn't done correctly. Be sure students understand what low impedance means, and how this relates to the need to connect the ammeter in series with the load.

9-6 Ammeter Shunts

Explain what an ammeter shunt is, how it works, and why it is used. Go through the process of how to calculate what size shunt might be required, and how to figure the values based on several different sized shunts. Refer students back to Ohm's law to compute the resistance.

9-7 Multirange Ammeters

Discuss what multirange ammeters are, and display one (with a shunt), if possible. Explain the importance of the various shunts, and the necessity for a shunt always to be connected to the meter. Explain what would happen if it was not connected.

Discuss the make-before-break-switch method of connecting shunts to a meter movement. Explain contact resistance.

9-8 The Ayrton Shunt

Explain how this is done, and why it is a better method than the make-before-break-switch method.

Calculating the Resistor Values for an Ayrton Shunt

Explain how to use the various formulas and when each would come into play. Have students work the examples in the text, and maybe a few that you put on the board. Explain how changing the selector switch setting changes the whole problem. Go carefully over Figures 9-24 through 9-27 with students during this explanation process.

9-9 AC Ammeters

Make sure students understand that shunts are used with AC ammeters to increase their range only, not to reduce it. Explain what a current transformer is, and why it is more readily used with an AC ammeter than shunts are.

Computing the Turns Ratio

Go over this process with students, checking for understanding. Have students write the formula (with appropriate symbols) in their notes for future reference. Make sure students remember some basic rules of equations:

*The object is get all the numbers on one side of the equation and all the variables on the other side.

*To achieve this, certain mathematical principles are used, and whatever is done to one side of an equation must also be done to the other side of the equation. So, if you multiply on one side to eliminate a mixture of numbers and variables, you must also multiply whatever is on the other side of the equation.

In the example on pages 216-217 of the text, after cross-multiplying, you are still left with a number value and a variable (unknown value) together. To isolate the variable on one side of the equation, each side of the equation has to be multiplied by 10. That gives you $1N_s$, which actually means N_s, and gives you the answer of 50 on the other side of the equation.

Current Transformers

Explain what a CT is and be sure students read the bold print warning on page 217, and understand what it means. Discuss the process of looping the wire in the transformer, and how this affects the amps needed to deflect the meter full scale.

9-10 Clamp-on Ammeters

Demonstrate one of these in class, or refer students to Figure 9-32 as you explain the use of it. Be sure to have students note that the number of times the wire is looped around the jaw of the ammeter, is the same number that the reading will have to be divided by to get the correct value.

9-11 DC-AC Clamp-on Ammeters

Make the distinction between AC ammeters, CTs, and DC-AC clamp-on ammeters. Explain the polarity process at work, and how that distinguishes these types from each other.

Discuss the Hall effect, a Hall generator, how it works, and the roll of the semiconductor in the overall process. Ask students if they can name situations in which semiconductors are used (for example, computers).

9-12 The Ohmmeter

Explain what the ohmmeter is used for, and discuss the two types. Be sure students understand that the meter must supply its own power in order to function.

The Series Ohmmeter

Go over the process of how the resistance is figured, how it is measured, and how the different settings are used. As always, having one to demonstrate in class is useful. Remind students that the meter must first be adjusted to zero before any measurements are taken, and readjusted back to zero when the scale is changed.

9-13 Shunt Type Ohmmeters

Discuss these and compare them with shunt ammeters. Remind students these are used to measure low levels of resistance.

9-14 Digital Meters

Digital Ohmmeters

Discuss how these differ from analog meters and caution students to look for metric numeric value indicators on the readout screen. Also, point out how they function differently than analog meters do. Caution students that no ohmmeter should ever be connected to a current when the power is on.

Digital Multimeters

Describe the workings of the digital multimeter. Explain how the instrument probably saves time by displaying the measurement being sought after, rather than requiring the operator to read a pointer on a scale. Also, explain that some have a range setting switch, and others do not.

Describe autoranging meters and how they differ from those with fixed range settings. Discuss the fact that digital meters do not have a loading effect. Also, have students note the low impedance level, and what that means.

9-15 The Low-Impedance Voltage Tester

Explain what this type of tester is used for, emphasizing that it should never be used to test low power circuits. Review the information on low impedance, and add to that the effect of having the grounded object. Note that it is not connected to the circuit, but to the voltmeter.

9-16 The Oscilloscope

Compare the oscilloscope to a VOM, and explain why the oscilloscope is necessary in situations in which a VOM would be useless. If you do not have access to an oscilloscope, refer students to Figures 9-49 and 9-50. Be sure students understand that an oscilloscope is a voltmeter that measures voltage over a designated period of time, giving a two-dimensional image of that measurement.

Voltage Range Selection

To explain this and time base, draw a grid on the board, and place the X and Y axis where they should go, making sure students know which one will be used to read range, and which one will be used to read time. Explain how to read the range value on the screen, and that both negative and positive voltage can by displayed.

The Time Base

Explain how this works, and repeat that time will be shown on the X axis. If you have an oscilloscope in the room, demonstrate the sweep per division that takes place. Change the adjustment, and demonstrate this again. In the absence of an oscilloscope, after explaining this, have students explain it back to you.

Measuring Frequency

Go over the formula for computing frequency. If an oscilloscope is available, have students set the voltage range at different settings and count the seconds it takes for one wave form to complete a cycle. Have them plug these numbers into the formula, and figure out the frequency at these various settings.

Attenuated Probes

Discuss these probes, what they do, and how they are used. Be sure students note that the compensation on these probes should be checked frequently.

Oscilloscope Controls

Again, an actual oscilloscope is an invaluable teacher's aid. If you have one, point out the various controls, and what their functions are. If you don't have an oscilloscope, have students study Figure 9-55.

Interpreting Wave Forms

Using an oscilloscope, or Figures 9-56 through 9-58, have students identify the alternating current in Figure 9-56, and explain how to identify whether it is AC or DC. Go through the same process using Figure 9-57, and identifying DC. Do the same with Figure 9-58. If you have an oscilloscope, allow students to practice working with it, using the probe, and interpreting the wave forms. Create some sample wave forms to show on the overhead, or to pass out in class, for students to interpret.

9-17 The Wattmeter

Explain the use of a wattmeter and the double magnetic field involved. Discuss how this allows the wattmeter to measure an AC or DC circuit.

9-18 Recording Meters

Ask students if any of them has any experience with any type of recording meters. Discuss how they work, and try to have one or two of the different types of charts available for student observance. Compare a recording meter to, for example, an EKG machine or a lie detector machine. Both record spikes and other irregularities based on heart rate and function (sort of the body's electrical system).

9-19 Bridge Circuits

Explain how a bridge circuit is formed into a parallel-series circuit. Explain what a capacitor is, as well as what an inductor is. Discuss a Wheatsone bridge and have students note that it is used for measuring resistance.

Go through the steps involved with solving for R_v and R_x using a bridge circuit. Make sure students understand how this is done.

Unit Round Up

This unit covered several new devices, so a good review is important. Go over the summary together in class. Assign the Review and Measuring Instruments Practice Problems as homework, to see if students can use what they have learned on their own. If they can't, go back over any weak areas until students are comfortable with their ability to use these devices, read them, and calculate any needed measurements.

UNIT 10
USING WIRE TABLES AND
DETERMINING CONDUCTOR SIZES

OUTLINE

10-1 Using the *National Electical Code* Charts
10-2 Factors That Determine Ampacity
10-3 Correction Factors
10-4 Computing Conductor Sizes and Resistance
10-5 Computing Voltage Drop
10-6 Parallel Conductors
10-7 Testing Wire Installations

KEY TERMS

Ambient air temperature
American Wire Gauge (AWG)
Ampacities (current carrying ability)
Circular mil
Correction factor

Damp locations
Dry locations
Insulation
Maximum operating temperature
MEGGER
Mil foot

National Electrical Code
Parallel conductors
Voltage drop
Wet locations

Anticipatory Set

Go over the unit objectives with the class, and alert them to key terms that they will need to learn. Have the most up to date code charts copied for each student or enlarged and on a wall. In addition, a small, but varied collection of insulated wires available for students to look at is of great benefit for the students. As always, having the real item (in this case the MEGGER) on hand for student practice is invaluable as a teacher's aid.

10-1 Using the *National Electrical Code* Charts

Carefully introduce the charts mentioned and ask students if any of them has ever used these charts. How involved your discussion of the charts needs to be will depend on how many students have used the charts before.

10-2 Factors That Determine Ampacity

Conductor Material

Discuss the selection of the right type of wire and why copper is used in all current housing. Ask if anyone has had to replace any old aluminum wiring with copper in an older house or business. Double check just to be sure that students understand that the wire *is* the conductor. You never know!

Insulation Type

Discuss the heat factor as it applies to insulation. Explain how certain insulations could be insufficient for the job needed. Work with the insulation chart and be sure students understand how to read **and** use this chart. The text refers to Table 310-13, which is not in the text due to its length. You will need to supply this Table in class for students to look over. If it is not too costly, providing a copy of this Table for each student is a good idea. You could charge students for the cost of reproduction. As students look over each Table, have them explain what each column is for.

10-3 Correction Factors

Have students look over Table 310-16. Make sure they see the ampacity correction factors at the bottom. Once students are familiar with the Table, have them do some practice problems solving for the ampacity. Be sure students realize that the column they found the ampacity in is the same column that the corrections factors are in, at the bottom.

More Than Three Conductors in a Raceway

Double check for student knowledge levels on terms such as raceway, conduit, dissipation, and conductors. Discuss various types of conduit and where and when they might be used.

Discuss the heat dissipation that occurs when more than three conductors are placed in a raceway. Have students practice solving for the maximum ampacities of a variety of conductor combinations, using Note 8a of the table, as shown in Figure 10-5.

Duct Banks

Go over how duct banks are used, and ask if any of the students has any experience with duct banks. Give them an opportunity to share their knowledge with the class. Ask why using duct banks is a good way to handle a large number of conductors.

10-4 Computing Conductor Sizes and Resistance

Have students copy the formula on page 262, into their notes, as well as what each symbol represents. They may not use it much, but if you are going to require them to know it on a test, then you need to spend some time discussing the use of this formula.

Long Wire Lengths

Explain when it is necessary to compute a wire size instead of using one of the Tables in the code. Go over the four steps involved in this process. In doing so, try to have a sampling of wire lengths for students to work with while going through the steps.

Discuss what a circular mil is and a mil foot. Go over how to use each in the process of computing wire size and resistance. Direct students to the chart on page 265, and give them a minute or two to go over it.

Computing Resistance

Walk students through the symbols and the formula for computing the resistance. Give students some practice problems to work on. Continue to work on these types of problems until students are comfortable with them. Remind students that Ohm's law still has its place in the final solution.

Refer students to Figure 10-13, and explain the differences among the three conductors. After they have spent some time identifying the various aspects of Figure 10-13, walk them through the problem and solution phase of figuring the length of the conductors. Make sure students completely understand this process before moving forward.

10-5 Computing Voltage Drop

Discuss voltage drop and go over the formula. Have students close the book while you write the problem on the board. Then, ask students to explain how to figure the solution. Next, give them Example 11 on page 270 to figure out. If necessary, walk them through the two solutions.

10-6 Parallel Conductors

Begin by explaining the reasoning behind using parallel conductors. Go over the five conditions that must be met, from the *National Electrical Code*, and have students write them in their notes.

Also explain the problem that could occur by the creation of a magnetic field with one large conductor in a metal conduit. Be sure that students understand what eddy currents are and why they can cause a heat problem, resulting in the insulation melting. Go over the added problem of hysteresis loss that occurs with conduits made of magnetic metals.

10-7 Testing Wire Installations

Explain the reason for testing the wire installation, and for using the MEGGER to do this. If you do not have either the hand crank model or the battery operated model available for students to see or use, then refer students to the pictures in Figures 10-21 and 10-22. Explain the process of using this device to test the wire installation. Be sure students understand the difference in checking for shorts versus checking for grounds.

Unit Round Up

Do the summary together in class and go back over the Tables covered in this unit. Have students do the review and practice exercises on their own, and then go over them together in class so you will know whether you need to go over any area again to clear up any confusion. Make sure students can do all of the objectives that were set out at the beginning of the unit. Review key terms.

SECTION 4
Small Sources of Electricity

UNIT 11
CONDUCTION IN LIQUIDS AND GASES

OUTLINE

KEY TERMS

Acids
Alkalies
Anode
Arc
Cathode
Copper sulfate

Cuprous cyanide
Electrolysis
Electrolytes
Electron impact
Electroplating
Ionization potential

Ion
Metallic salt
Sulfuric acid
X-rays

Anticipatory Set

Go over the unit objectives and the key terms that students will be expected to know. Review valence electrons, referring students to Unit 1, Section 6, of the text. Draw a few atoms showing the valence shell and the valence electrons on the board, or refer students to Figures 11-1 and 11-2 in the text.

Ask students for examples (such as different types of batteries) of things that conduct electrical current in gas or liquid.

11-1 The Ionization Process: Magnesium and Chlorine

Discuss positive and negative ions. Be sure that students understand the process in which magnesium gives its two valence electrons to the chlorine atom, making the magnesium a positive ion, and the chlorine a negative one. Continue with this, explaining why adding salt to water changes it from an insulator to a conductor. Discuss electrolytes and how and where they are used.

11-2 Other Types of Ions

Sulfuric Acid

Discuss how sulfuric acid is formed. Be sure students note the fact that the two hydrogen atoms gave their electrons to the sulfur-water combination, and that the resulting sulfate ion is actually held together by these added electrons.

Copper Sulfate

Explain how copper sulfate is formed and how it is used. Again, have students note the double plus sign attached to the cupric ion, denoting it as positive, and the double negative sign on the sulfate ion, denoting it as negative.

Cuprous Cyanide

Explain what this is and how it is used.

11-3 Electroplating

Ask students what they know about electroplating, and if they can name any examples. Have them enter the five factors of electroplating into their notes. Make sure students make the distinction between the cathode and the anode. Explain the electroplating process and why it is used.

11-4 Electrolysis

Explain this process and why it is used. Compare it with electroplating, and make sure that students can explain the differences between the two processes.

11-5 Conduction in Gases

Discuss Figure 11-6, pointing out the electrodes and the air gap. Explain the relationship of the distance between the electrodes and the amount of voltage needed to initiate the arc. Make sure that students understand why the same amount of voltage is not needed to maintain the same arc.

Ionization Potential in Gas

Define ionization potential and then discuss each factor that determines the amount of voltage required to cause conduction in a gas-filled envelope. Cover each factor, having students explain each factor's individual effect on the required amount of voltage.

Electron Impact

Explain how important electron impact is to the ionization of a gas. Also discuss how the density of the gas in the tube or envelope affects the speed at which this ionization process will take place. Refer students to Figures 11-9 through 11-11 as you explain this.

Useful Applications

Go over the common uses of ionization in a gas. Ask students if they can think of any other examples of this process in action.

11-6 Ionization in Nature

Discuss the phenomena caused by ionization of gas in the atmosphere. Ask students for other examples (swamp gas).

Unit Round Up

Go over the summary in class, and assign the review questions for homework. Clear up any confusion that shows up as you go over the review questions during the next class session. Go back over the chapter objectives to make sure they have been met, and review all key terms.

UNIT 12
BATTERIES AND
OTHER SOURCES OF ELECTRICITY

OUTLINE

KEY TERMS

Battery
Cell
Current capacity
Electromotive series of metals
Hydrometer

Internal resistance
Load test
Nickel-cadmium (ni-cad) cell
Piezoelectricity
Primary cell

Secondary cell
Specific gravity
Thermocouple
Voltaic cell
Voltaic pile

Anticipatory Set

Ask students what experiences they have had with batteries. Have a variety of batteries on display. As you go through each section, hold up or pass around the various types of batteries that you are discussing.

Another great visual aid is a voltmeter hooked up to a potato with aluminum and copper wires, as well as a voltmeter hooked up to a nickel and a penny, as in Figure 12-4. The more visual aids you have, the easier the learning experience becomes.

12-1 History of the Battery

Briefly go over the history of the battery. Ask students to explain why the saline solution on the frog caused the frog to become a conductor of electricity.

12-2 Cells

Explain what a voltaic cell is and describe how easily one can be constructed. If you have the potato, the voltmeter, and the wires needed, hook up this cell for students to see. Explain that the voltage produced is quite small. After describing the cell created in Figure 12-4, ask students if they have ever licked the top of a 9 volt battery to see if it was still good. Then, ask them if they now understand why they received a slight shock on the tongue (acid or alkali in their saliva acting as an electrolyte).

12-3 Cell Voltage

Explain that the type of materials used to make a cell is what determines its voltage. Refer students to the charts of metals in Figures 12-5 and 12-6. Explain the difference between a primary cell and a secondary cell.

12-4 Primary Cells

Discuss the process taking place between the zinc and copper suspended in the conductive solution. Have students explain to *you* what is taking place.

The Carbon-Zinc Cell

Discuss the process Leclanché came up with. Have students explain why this combination of materials worked. Be sure they know what a dry cell battery is.

Alkaline Cells

Describe alkaline cells and compare them to carbon-zinc cells. Discuss the advantages and disadvantages of alkaline cells.

Button Cells

Discuss various types and have students give examples of items that they use button cells in. Explain the differences between mercury-zinc button cells and silver-zinc button cells. Refer students to Figure 12-13 for an excellent view of each part of the mercury button cell.

Lithium Cells

Describe the various types of lithium cells. Discuss why different types of cathode materials can be used, as well as different types of electrolyte substances. Be sure students make note of the shelf life of those using a solid electrolyte. Ask students why that might be important.

Current Capacity and Cell Ratings

Have students write the definition of current capacity into their notes, as well as the statement that the amount of current a particular type of cell can deliver is determined by the surface area of its plates. Hold up a size D battery next to a size AA or AAA battery to illustrate the point.

Discuss the factors necessary to determine a cell's current capacity. Explain what the milliampere ($_mA$) is, and how it applies to rating primary cells. Go through the process of figuring watt-hours, and watt-hours per cubic inch.

Internal Resistance

Discuss that all cells have internal resistance and how this is affected as the cell is used and ages. Have students explain to you what this means in terms of real milliamperes delivered over a period of time.

12-5 Secondary Cells: Lead-Acid Batteries

Explain what a secondary cell is and ask students for examples. Define specific gravity and explain what a hydrometer is.

Discharge Cycle

Discuss this cycle and have students explain how the hydrometer can be used to check for the level of charge left in the battery.

Charging Cycle

Explain this process and have students give examples of ways this can be done. Also, ask students why they think people are always warned not to smoke around a battery being jump charged.

Cautions Concerning Charging

In addition to the warning about smoking around a charging battery, be sure students are aware of the other hazards involved with overcharging a battery.

Sealed Lead-Acid Batteries

Describe these types of batteries, referring students to Figure 12-24. Ask students why these types of batteries can be more advantageous than those using a liquid electrolyte.

Ratings for Lead-Acid Batteries

Discuss amp-hour ratings and how the rating is determined. Also discuss cold cranking amps for automotive batteries.

Testing Lead-Acid Batteries

Explain the load test and why it might need to be used when the hydrometer reading cannot be trusted. Discuss how the test is conducted and what voltage level should be maintained for the specified amount of time.

12-6 Other Secondary Cells

Nickel-Iron Batteries (Edison Battery)

Describe these batteries and discuss their advantages and disadvantages. Discuss in what circumstances they would be more useful than a lead-acid battery.

Nickel-Cadium (ni-cad) Batteries

Describe these batteries and discuss their advantages and disadvantages. Explain the discharge-recharge memory factor of this battery. Ask students if they have used any batteries like this (most camcorder batteries are this type of rechargeable battery with a memory curve).

12-7 Series and Parallel Battery Connections

Explain the effect of connecting batteries or cells in series. Make sure students understand that the voltages add, while the current capacities remain the same. Refer students to Figures 12-27 through 12-29. Emphasize that batteries of different voltages should never be connected in parallel. Go over how batteries can also be connected in a series-parallel combination.

12-8 Other Small Sources of Electricity

Solar Cells

Explain what solar cells are and how P-type material and N-type material are formed. Discuss what happens when the photons of light strike the surface of the photo cell. Explain the necessity of creating connections of solar cells in series or parallel or both in order to obtain desired amounts of voltage and current.

Thermocouples

Explain the Seebeck effect, and a thermocouple. Go over what determines the amount of voltage the thermocouple can produce. Have students read over the chart in Figure 12-34 of the most common types of thermocouples.

Discuss the need to connect thermocouples in series, as a thermopile. Ask students if they are familiar with thermocouples that exist where they live.

Piezoelectricity

Explain what piezoelectricity is and how it is produced through pressure. Compare this to the thermocouple that uses heat to produce voltage. Discuss various places in which piezoelectricity is used, especially where heat would be an inappropriate way of producing voltage. Discuss the various types of crystals that are used to produce piezoelectricity.

Unit Round Up

Go over the summary together in class and have students do the review exercises for homework. Discuss those problems during your next class session, checking for understanding as you do. Go over the unit objectives to be sure that they have been met, and review key terms.

UNIT 13
MAGNETIC INDUCTION

OUTLINE

KEY TERMS

Eddy current
Exponential curve
Henry (H)
Hysteresis loss
Magnetic induction

Lenz's law
Metal oxide varistor (MOV)
R-L time constant
Speed
Strength of magnetic field

Turns of wire
Voltage spike
Weber (Wb)

Anticipatory Set

Go over the objectives for this unit and key terms. A review of magnets and the role they play in electric currents (Unit 4) is a good place to start.

Visual aids may be difficult to come by for this unit, but any electromagnetic items that you can display will be helpful.

13-1 Magnetic Induction

Go over the principle of magnetic induction and be sure students fully understand what it means. Discuss how the polarity changes, and have students write the note about the polarity of the induced voltage into their notes. Check for understanding before moving to the next section.

13-2 Fleming's Left-Hand Generator Rule

Demonstrate this rule using your own left hand, and have students do the same. Explain why this is a helpful way to remember this principle in the field.

13-3 Moving Magnetic Fields

Go over the three major components of magnetic induction. Also, explain why it is more desirable to move the magnet, instead of the conductor, to produce the lines of flux and induce a voltage.

13-4 Determining the Amount of Induced Voltage

Have students write in their notes the three factors that determine the amount of voltage that will be induced in a conductor. Explain the process of increasing the amount of voltage in each conductor by increasing the Wb/s.

13-5 Lenz's Law

Go over Lenz's law, and have students write it into their notes. Make sure that students understand how this affects polarity. Be sure that students understand what EMF and CEMF represent. From this discussion, move into an explanation of why inductors always oppose a change of polarity, and how this affects induced voltage.

13-6 Rise Time of Current in an Inductor

Discuss the rise time of current in an inductor, and compare it to the rise time of current through a resistor. Have students think of examples for each type of rise in current. Also, make sure students understand that the induced voltage has the opposite polarity of the applied voltage, and that induced voltage is proportional to the rate of change of current. Refer students to Figures 13-13 through 13-15, as well as any examples you could draw on the board or overhead.

13-7 The Exponential Curve

Explain the concept of using the exponential curve to measure the rate of occurrences of some variable, such as amps, water, etc. Draw this on the board, or have a container of water in the room that you can release water from. This will show the 63.2% release of water in one minute better than any picture can. An ice chest with a plug on the side will work just fine for this visual.

13-8 Inductance

Have students write the rule for inductance into their notes, making sure that they include what the henry (H) is, and that the symbol for inductance is L. Discuss air core inductors and iron core inductors, distinguishing them from each other by their different qualities. Go over eddy current and hysteresis loss, having students look back at the notes they took from Section 6 of Unit 10. Also, go over the difference in how tightly the coil is wrapped around the conductor, resulting in the windings being spaced closer or farther apart, and what that means.

13-9 R-L Time Constants

Explain what an R-L time constant is, and go over the formula for computing the time of one time constant. Relate this back to the exponential curve, and the need to know what each time constant is. Ask students why it is necessary to know the time constant, and what would happen if they did not know it. Also, go over the formula for inductance and have students add this to their notes.

13-10 Induced Voltage Spikes

Describe what causes a voltage spike, and go through the process of what happens when the switch is open (series circuit composed of the resistor and the inductor), and when the switch is closed (current flowing through the inductor). Go over Figures 13-20 through 13-23 carefully as you explain this.

Explain what the diode is and how it prevents induced voltage spikes. Also, describe how the diode is hooked up with the inductor, and that this makes the diode reverse-biased, like a closed switch. Make sure students note that the diode can only be used this way in direct current.

Describe the metal oxide varistor (**MOV**), and explain how it can prevent induced voltage spikes in either direct **or** alternating current. Utilize Figures 13-24 through 13-26 during this discussion, to help students understand how these things work. Explain where an MOV would be used, and why.

Describe the electric fence charger, and explain how it works. Compare it to the MOV, pointing out key differences. Ask when an electric fence charger would be used.

Unit Round Up

Check for understanding by going over the summary in class and having students explain each process. Have students describe the various items used to prevent induced voltage spikes, and double check to be sure that they can distinguish between each one. Assign the review questions for homework, and go over them together during the next class session. Review the unit objectives and key terms.

SECTION 5
Basics of Alternating Current

UNIT 14
BASIC TRIGONOMETRY

OUTLINE

KEY TERMS

Adjacent	Oscar Had A Heap Of Apples	Right triangle
Cosine	Pythagoras	Sine
Hypotenuse	Pythagorean theorem	Tangent
Opposite	Right angle	

Anticipatory Set

This unit will prepare students for the math functions they will need to be able to use throughout the rest of the text. With this in mind, create practice sheets for students to work on every day. As in all areas of math, practice is the best way to understand how a formula works and how to identify the variables. Practicing the functions learned in this unit will allow students an opportunity to gain confidence in using these functions comfortably. In addition, students will benefit greatly from being able to work these problems on the chalk board. This type of exercise allows you to see where an individual misses a key step, or lacks understanding of how something is supposed to be done. Being able to pinpoint the exact place where confusion starts will enable you to help your students better. As always, review the unit objectives, and all key terms.

14-1 Right Triangles

Explain what a right triangle is and draw one on the board. Label the hypotenuse and make sure students understand that it is the longest of the three sides of the triangle. Also, be sure that students realize that not every triangle is a right triangle. Make sure they understand that the presence of the $90°$ angle is what makes a triangle a right triangle.

14-2 The Pythagorean Theorem

Explain how the theorem works and give students some practice problems. Remind students that they are working with squared numbers, but that the answer will be the *square root* of the sum of the two given factors. Also, show students how this will work with any missing factor, as long as any two factors are available. For example, $A^2 = C^2 - B^2$; $B^2 = C^2 - A^2$; and $C^2 = A^2 + B^2$.

14-3 Sines, Cosines, and Tangents

Draw a right triangle on the board and show students the $90°$ right angle. Explain that the sum of the other two angles must also be 90^0, because the total sum of the three angles of *any* triangle is $180°$ ($180° - 90° = 90°$).

Using the right triangle you have already drawn, bisect it and show how side A is opposite angle X, but adjacent to angle Y. Show how side B is opposite angle Y, but adjacent to angle X. Have students write the sine function into their notes and on a flash card. In fact, have them each make one flash card with Oscar Had A Heap Of Apples on it, as is shown on page 368.

Once students have looked over these formulas, give them some practice right triangles to use to solve for missing angles. Remind them that once they have the lengths of any two sides, they can use the Pythagorean theorem to solve for the third side. Once they have the third side length, they can solve for the sine, cosine, and the tangent.

If students have a scientific calculator (and it would be a good idea for them to get one), show them, using the calculator, how to determine the angle after solving for the sine. If they do not have the calculator, refer them to Appendix A in the back of the text. Be sure they realize that when looking for the sine, they will use the left-hand column, which is continued on the next page of the Appendix.

Once they have one of the angles solved, show them that they can subtract the newly found angle from $90°$ to obtain the remaining angle ($90° - 40° = 50°$). They could also use the cosine formula.

Remind students that if they use the Pythagorean formula when they have two known side lengths, they can easily solve for the sine, establish the angle, subtract that angle from $90°$, and determine the third angle.

14-4 Formulas

Have students copy these formulas into their notes. It is a good idea to do flash cards on these as well. Take some time here to practice using each of these formulas.

14-5 Practical Application

Have students practice solving for the needed information in the example in the text. Propose some other problems for them to solve, using situations that they might encounter in the field.

Unit Round Up

Go over the summary in class, and assign the review questions and the practice problems for homework. Check for understanding as you go over the homework during your next class session. Reteach any areas where there is still confusion. Review the unit objectives and key terms.

UNIT 15
ALTERNATING CURRENT

OUTLINE

KEY TERMS

Amplitude   In phase   RMS
Average   Linear wave   Sine wave
Cycle   Oscillators   Skin effect
Effective   Peak   Triangle wave
Frequency   Resistive loads   True Power
Hertz(Hz)   Ripple   Watts

Anticipatory Set

Providing a short review of the previous unit might be necessary, as that information will be put to use in this unit. Be prepared to take whatever time is necessary for students to feel comfortable solving for voltage values. Be prepared to do a lot of work at the chalkboard or on the overhead.

Practice sheets for solving voltage values are a good warm-up exercise. Go over the unit objectives and key terms.

15-1 Advantages of Alternating Current

Go over the advantages of AC over DC, and be sure students realize that the greatest advantage is lower cost.

15-2 AC Wave Forms

Square Waves

Explain what square waves are and how they are formed by the constantly changing polarity of the voltage. Refer students to Figure 15-1. Define what an oscillator is, and explain how it is used to change DC voltage into AC voltage.

Triangle Waves

Explain what these are and point out the linear lines that show these waves in Figure 15-5. Make the distinction between triangle waves and square waves.

Sine Waves

Explain that this is the most common wave form they will encounter, and how they are formed. Explain what a cycle is and how this determines frequency. Be sure students understand what a hertz (Hz) is and that the most common frequency in the U. S. and Canada is 60 Hz. Have students copy the rule about sine waves into their notes.

Refer to Figure 15-7 as you explain the process of the loop cutting through the flux lines, and how the angle of that cut affects the induced voltage. Instruct students to copy the formula for the instantaneous voltage value into their notes. Do the practice problem on the board. Have students refer to their notes from the previous unit if they need to. They can either use calculators to find the SIN of 78°, or Appendix A in the back of the text. Once they have the correct figures plugged into the formula, they can proceed to a solution.

15-3 Sine Wave Values

Refer students to the glossary for exact definitions of the value terms listed. Then illustrate them by drawing them on the chalk board. As you move through the sub-sections on each of these terms, ask students to explain each of them to you. Also, utilize the figures in the text for further illustration.

Peak-to-Peak and Peak Values

Have students explain these to you, as they are very basic terms.

RMS Values

Discuss RMS values at length, and have students write the definition and the formula for RMS in their notes. They also need to copy the formula for peak. Explain how the RMS value is used, and be sure that students realize that throughout the remainder of this text, all values of AC voltage and current will be RMS values.

Average Values

Explain that these are really DC values. Go through the steps for solving for an AC sine wave DC voltage value. Be sure students add the new formula to their notes, as it is quicker than going through peak to find the average. Then, expand this discussion by adding the half-wave rectifier, and modifying the full wave rectification formula to solve for the half-wave DC voltage average.

Wavelength

Explain what wavelength is and demonstrate how to use the formula to compute a wavelength. Refer to Figures 15-17 and 15-18 as you describe this process.

15-4 Resistive Loads

Explain what resistive loads in an AC current are and have students take note of the two main characteristics of resistive loads. Remind students that any time a circuit contains resistance, electrical energy will be changed into heat. Explain what it means for the current to be *in phase* with the voltage.

15-5 Power in an AC Circuit

Discuss what it means to say that true power, or watts, can be produced only when both current and voltage are either positive or negative, and how this results in wattage always being positive.

15-6 Skin Effect in AC Circuits

Discuss what this is and how it affects resistance, but more importantly, how it is affected by frequency levels. Explain that the problem of skin effect in high frequency situations is dealt with by using conductors with a larger surface area.

Unit Round Up

Go over the summary in class, and assign the review and practice problems for homework. Allow students to do the homework problems on the board during the next class session. See if they can explain what they did, and why they chose the method they used. Review the objectives and key terms of this unit.

SECTION 6
Alternating Current
Circuits Containing Inductance

UNIT 16
INDUCTANCE IN
ALTERNATING CURRENT CIRCUITS

OUTLINE

KEY TERMS

Current lags voltage

Impedance (Z)

Induced voltage

Inductance (L)

Inductive reactance (X_L)

Quality (Q)

Reactance

Reactive power (VARs)

Anticipatory Set

A short review of Unit 13, especially as inductance is concerned, is a good idea. Remind students of the effect of magnetic lines of flux, the opposite polarity involved with induced voltage, the proportion of reduced voltage to the rate of change of the current, and an inductor's opposition to a change of current.

16-1 Inductance

Describe what inductance is and make sure students understand that there will always be a small amount of inductance present in all alternating currents. Emphasize that circuits in items such as motors, transformers, and chokes all contain coils, so inductance is a factor.

Explain the process of the applied voltage overcoming the induced voltage in order for current to flow through the circuit.

Computing the Induced Voltage

Discuss how to compute induced voltage and remind students they will use Ohm's law.

16-2 Inductive Reactance

Define reactance and explain how it occurs, and move into a discussion of inductive reactance (X_L). Have students copy the formula for X_L into their notes. Go over the three factors that inductive reactance is proportional to. All of these things have already been covered in previous units. Work through with the students the three practice problems given in the text.

16-3 Schematic Symbols

Have students write the iron core and air core symbols for induction into their notes.

16-4 Inductors Connected in Series

Explain how to determine the L_T and X_{LT} of a circuit. It is similar to finding resistance.

16-5 Inductors Connected in Parallel

Provide a brief review of solving for resistance in parallel circuits (Unit 7). Then go through the formula for solving for L_T in a parallel circuit.

16-6 Voltage and Current Relationships in an Inductive Circuit

Go through this slowly, drawing the wave on the board as you explain the $90°$ lag of current behind voltage. Point out the peak point, as well as the current passing from negative, through zero, to a positive direction. As you explain this, make students aware of the relationship of the magnetic field: it expands and collapses with this process.

16-7 Power in an Inductive Circuit

Explain the power phases as polarity changes and the magnetic field collapses. Again, draw this on the board, and have students tell you what is occurring when the current and voltage both have the same polarity, and when they don't. Be sure students understand the relationship of this polarity to the magnetic field. Go over types of losses as they relate to power used, and make sure students understand the expression *true power*.

16-8 Reactive Power

Explain what VARs is, and work through the formula for computing it. Compare VARs to watts, and note the similarities in computing VARs amounts.

16-9 Q of an Inductor

Explain what determines high and low quality in an inductor. Go through the formula to compute Q, and make sure students note that if Q=10 or more, the inductor is considered to be a pure conductor because the resistance becomes negligible.

Have students refer back to their trigonometry notes, if necessary, prior to explaining impedance and how to solve for it, using the pythagorean theorem.

Unit Round Up

Go over the summary and all key terms in class. Have students explain back to you the principles discussed as an initial check for understanding. Give the review questions and practice problems for homework. Check these together, in class, before advancing to Unit 17.

UNIT 17
RESISTIVE-INDUCTIVE SERIES CIRCUITS

OUTLINE

KEY TERMS

Angle theta ($\angle\emptyset$)
Apparent power (VA)
Parallelogram method
Power factor (PF)
Quadrature power
Reactive power (VARs)
Total current (I)

Total impedance (Z)
True power (P)
Vector
Vector addition
Voltage drop across the inductor (E_L)
Voltage drop across the resistor (E_R)
Wattless power

Anticipatory Set

Begin by reviewing the relationships that are out of phase with each other, and how that shows up on the screen. Refer to Figure 17-1 to emphasize this. Also, provide a short review of Unit 14, and right triangles. This will help set up the sections that use vector addition as a problem solving tool. Make sure students can identify a right triangle, a hypotenuse, and how to find the hypotenuse.

17-1 R-L Series Circuits

Explain how a pure resistive load and alternating current circuit become in phase with each other. Discuss how to figure the exact amount of phase angle difference. Have students copy the list of symbols and corresponding definitions into their notes.

Go over the example in Figure 17-2 on the board. Have students explain how to figure for the missing quantities. Remind them of formulas they can use with the information given.

17-2 Impedance

Review the definition of impedance, and discuss combining it with inductive reactance. Be sure students understand that in resistive-inductive series circuits, total impedance (Z) will be the sum of the resistance and the inductive reactance. Go over the formula to solve for Z. If necessary, do a brief review of the situation in which two quantities are $90°$ out of phase with each other and form a right triangle (Unit 14).

17-3 Vectors

Distinguish between vectors and scalars. Double check for understanding by having students give examples of each, similar to those given in the text. Be sure students understand that subsequent vectors are drawn from the reference point of $0°$, not from any other previously drawn vector. Also, make sure that students note the counterclockwise movement of vectors.

Adding Vectors

Explain that for electrical purposes vectors will only be added, and that the method used to accomplish this is called vector addition. Go over the process of adding vectors that are going in the same direction.

Adding Vectors with Opposite Directions

Explain that to be opposite, there must be a $180°$ difference in the directions of the vectors. Go over the simple algebra equation used to solve for the resulting magnitude.

Adding Vectors of Different Directions

Using Figures 17-10 and 17-11, explain how to add vectors by connecting starting and ending points. Emphasize that resistance and inductive reactance are $90°$ out of phase with each other, and that the two vectors must be placed at a $90°$ angle. This is shown clearly in Figure 17-11. With this information, have students explain how to find the impedance of the circuit.

The Parallelogram Method of Vector Addition

Explain what a parallelogram is and how one can be formed by two vectors originating from the same point. Discuss why this is such a good method for determining the resultant of two vectors that originate from the same point. Go through the various ways the parallelogram can be used when dealing with vectors.

17-4 Total Current

Explain how to replace R in an Ohm's law formula with impedance to solve for total current (I_T). Practice this, using the example in Figure 17-17.

17-5 Voltage Drop across the Resistor

Explain how the voltage drop is computed, and emphasize that these quantities cannot be found using Ohm's law. Explain how these quantities can be found, if they are like resistive quantities, using vector addition. Before moving on, be sure students understand the formulas they can and can't use to solve for the amount of voltage dropped across the resistor. If necessary, go over this again.

17-6 Watts

Remind students that watts can only be produced when voltage and current are both positive, or both negative. Ensure that students understand what formulas can be used to find the watts quantity, and what formulas cannot be used. Clear up any confusion before going to the next section.

17-7 Computing the Inductance

Go over this formula, and have students work practice problems on the board.

17-8 Voltage Drop across the Inductor

Go over this formula, and have students practice at the board.

17-9 Total Voltage

Explain that the total voltage in R-L series circuits is the sum of the voltage drops, just as it is in other series circuits. Go over the ways the total voltage can be solved for, allowing students time to practice each one. Explain when one formula might be better to use than another.

17-10 Computing the Reactive Power

Explain what reactive power (VARs) is, and distinguish it from watts. Explain why it is also referred to as quadrature power or wattless power. Go over the formula to compute reactive power, and do some practice problems.

17-11 Computing the Apparent Power

Define apparent power and explain how it is computed. Go over the formula, and show students how VA can be figured using vector addition. Do some practice problems using both methods to make sure students understand the process.

17-12 Power Factor

Explain what PF is and how easily it can be determined. Explain that the PF is usually given as a percentage. Make sure that students make note of the fact that PF cannot be figured in a series circuit using current, because the current is the same all through a series circuit. Explain why the PF amount is important, especially in a business or industrial setting.

17-13 Angle Theta

Define angle theta and explain the two ways it can be solved for. Draw the parallelogram and show students the vector addition method. Refer to Figure 17-25 to demonstrate the cosine of angle theta as the PF. Check to be sure that students can differentiate between the resistive leg, the total leg, and identify where each is located on the triangle. Go over the trigonomic functions that students can also use to solve for angle theta.

Work through the sample circuit in Figure 17-29 and make sure students understand how to solve for the missing quantities. If there is any confusion, create another sample, and continue to practice this until students are comfortable with the various solution processes.

Apparent Power

Explain what apparent power is. Make sure students understand the difference between apparent power (measured in volt-amps) and true power (measured in watts). Go through the steps of solving for apparent power.

Total Circuit Current

Walk students through solving for the total circuit current, utilizing the applied voltage and Ohm's law.

Other Circuit Values

This last section is key, and practice will be extremely important. Have students go through the steps of solving for impedance, power factor, angle theta, and the other missing values. If confusion exists, do another practice circuit on the board. Have students go to the board to practice the various solution formulas. Also have them explain why they are doing what they are doing.

Unit Round Up

Go through the summary in class, and be ready to stop and reteach any area of uncertainty. Check to see if students are having problems with the circuit information, or if the problem is simply a math deficiency. Some students may need some extra math tutoring if the formulas are confusing them.

Assign the review questions for homework, but go over them during your next class session. Also have students complete the practice problems on their own (perhaps in class, where you can monitor their work). Before moving on to Unit 18, have students work their practice problems on the board while you watch for any problem areas.

UNIT 18
RESISTIVE-INDUCTIVE PARALLEL CIRCUITS

OUTLINE
18-1   Resistive-Inductive Parallel Circuits
18-2   Computing Circuit Values

KEY TERMS
Angle theta ($\angle\varnothing$)
Apparent power (VA)
Current flow through the inductor (I_L)

Current flow through the resistor (I_R)
Power factor (PF)
Reactive power (VARs)

Total current (I_T)
Total impedance (Z)
True power (P)
Watts

Anticipatvory Set
Go over the unit objectives and key terms with the students. This unit simply builds on what they learned in the previous unit. If any students had difficulty with Units 16 and/or 17, this bit of news will be encouraging to them.

As you go through the key terms, allow students to define the ones they know. This will be a confidence builder and an aid to absorbing the information in Unit 18.

18-1 Resistive-Inductive Parallel Circuits
Describe what a resistive-inductive parallel circuit consists of, and how the various components relate to each other. Discuss phase angle and what causes the amount of phase angle to be different, based on the ratio of the amount of resistance to the amount of inductance in the circuit. Also, make sure students realize that solving for the circuit power factor doesn't change.

18-2 Computing Circuit Values
Using Figure 18-2, be sure students understand the lag element of the current flow through the inductor, as it relates to the voltage. Go over each part of the circuit listed in the text to ensure that students are aware of what they will be solving for in the circuit in Figure 18-4.

Resistive Current
Remind students that voltage remains the same across each component of the circuit. Have them solve for the current flow through the resistor using $I_r = E/R$.

Watts
Remind students of the definition of what a watt is, and review previously learned formulas. Go through this formula, checking for understanding.

Inductive Current
Make sure students understand that they can determine the current flow through the inductor, because they know what voltage is being applied to the inductor (240).

VARs
Go over this formula with students, making sure they understand that the correct formula is current flow through the inductor, multiplied by the voltage applied to the inductor. Again, once the applied voltage is known and the inductive current is found (using the applied voltage), then solving for VARs is a simple multiplication problem. Any time you can break something down to its simplest form, it helps build student confidence.

Inductance
Walk students through this formula, making sure they know what they are dividing.

Total Current

Explain that total current is the current flow through the resistor added to the current flow through the inductor. Students, however, need to keep in mind that the two currents are $90°$ out of phase with each other, making it necessary to use vector addition. If necessary, do a quick review of vector addition.

Impedance

Walk students through this simple Ohm's law-type formula, checking for understanding as you go. Explain how the value of impedance can also be found when the total current and voltage are not known. Go through this process on the board, making sure that students are following the process. Have a few students come up and do a practice problem, solving for impedance without having the total current and voltage amounts.

Apparent Power

Go through the steps for computing apparent power. Explain that true power and reactive power form a right triangle, because they are $90°$ out of phase with each other. Point out that the hypotenuse of this right triangle is the apparent power.

Power Factor

Explain that PF is the same in an R-L parallel circuit, and an R-L series circuit but the formulas used to compute these values are different. Go through these formulas with your students, and make sure they understand the differences between these formulas and those used for series circuits.

Angle Theta

Using Figure 18-8, show students the angle produced by the apparent power and the true power. Make sure students understand how to determine the cosine of angle theta.

Have students do the example in Figure 18-11, in class, so that you can spot any areas of confusion or weakness.

Then, go step by step through the solution process with the students, discussing any difficulties they may have had solving for the various amounts. These discussions will help you see where a confused student loses the thread of a certain process. Since all of these processes build upon one another, it is important to follow students' thinking through the process of solving a problem, to see where they stop understanding and begin improvising.

Unit Round Up

Go over the summary in class, and assign review questions and practice problems for homework. Discuss the review question answers during your next class session, and have students work the practice problems on the board and explain how they arrived at their answers so that you can check for understanding.

SECTION 7
Alternating Circuits
Containing Capacitance

UNIT 19
CAPACITORS

OUTLINE

KEY TERMS

Dielectric
Dielectric constant
Dielectric stress
Electrolytic
Electrostatic charge
Exponential

Farad
HIPOT
JAN standard
Leakage current
Nonpolarized capacitors
Plates

Polarized capacitors
RC time constant
Surface area
Variable capacitors

Anticipavtory Set

Go over the unit objectives and key terms with students and provide a short review on electrostatic charges (Unit 3). If possible, have a few resistors and capacitors on hand to show students. Allow students to handle both the capacitors and the resistors. It is important that they are able to distinguish between the two by sight. You may also need to do a review on polarity (Units 3 and 13). If possible, have at least one of the various types of capacitors and testing mechanisms to use at the end of the unit.

19-1 Capacitors

While holding up a capacitor or passing a few around the room, explain what a capacitor is. Be extremely firm when warning students against charging up a capacitor before handing it to someone else. Go over the three things that determine a capacitor's capacitance, or the electrical size of the capacitor.

Explain what a dielectric is, and review the different types of materials that are used as dielectrics.

Discuss surface area, and how that affects the capacitor and the movement of the electrons from one plate to another. Be sure students realize that once a capacitor has been charged, it stays charged until a path is introduced for the electrons to travel on. Point out the rule concerning capacitors and current flow.

Explain leakage current and how it occurs. Remind students that insulators are materials with seven or eight valence electrons, and are used because they resist the flow of electricity. Explain that a dielectric is a type of insulator, but that no dielectric is a perfect insulator. The result of this is leakage current, and the possibility of the capacitor becoming a leaky capacitor.

19-2 Electrostatic Charge

Explain how a capacitor stores energy, and differentiate between a capacitor's stored energy and the stored energy of an inductor.

Dielectric Stress

Explain what dielectric stress is and how it occurs. Refer to Figures 19-6 and 19-7 in the text to illustrate. Describe the potential, or stored energy effect that this stress has on the capacitor.

Discuss voltage as it relates to shorting out a capacitor and lowering the life span of a capacitor. Show students how to identify the voltage rating of a capacitor, and warn them, again, of the consequences of exceeding that rating.

Discuss how the energy of the capacitor is stored in the dielectric as an electrostatic charge. Using the examples from the text of the camera flash and the 0.357 cartridge, ask students if they understand why a charged capacitor should not be handed to someone. Ask about shocks students may have received after sliding across a car seat and touching the car door. Enlarge that for them, exponentially, to communicate the potential waiting to be released from the capacitor.

19-3 Dielectric Constants

Explain how this number is obtained and what it means. Make sure students know that a farad is the basic unit of capacitance.

19-4 Capacitor Ratings

Remind students of what a coulomb is. Explain that the change of a volt across the plates of a capacitor, resulting in the movement of 1 coulomb, gives the capacitor a capacitance of 1 farad.

Go over the chart of various dielectric materials, and the dielectric constant levels of those materials. Remind students that air is always the starting point. Ask students why this is the case, and see if they can relate it back to electrostatic charges.

Carefully go through the steps involved in solving for capacitance of a capacitor. Make sure students understand when to plug in micro, nano, and pico farads to make the problem easier to work with.

19-5 Capacitors Connected in Parallel

Discuss connecting capacitors in parallel and why this is done. Go through the simple procedure used to determine the total capacitance of capacitors set up in parallel.

19-6 Capacitors Connected in Series

Distinguish between capacitors set up in series and capacitors set up in parallel. Make sure students understand that a parallel system adds one capacitor to another, for an overall capacitance larger than any one capacitor in the parallel connection. The capacitors in series, however, are the opposite, reducing the overall capacitance when connected in series. Go over the formula used to solve for capacitance when connected in series.

19-7 Capacitive Charge and Discharge Rates

Make sure students understand the difference between changing exponentially versus added change. Two examples you can use are hurricanes and earthquakes. Explain that an earthquake registering a 5 on the Richter scale is not one level smaller than an earthquake registering a level 6. Additionally, a category 3 hurricane is not one level weaker than a category 4 hurricane. Make sure students understand the power potential that accumulates under exponential growth.

Show students, on the board, how the charging pattern occurs in a capacitor at a level of 63.2% of whatever the required charge is, over and over, until the maximum required charge has been reached. Explain that this is also true in reverse for discharging a capacitor.

19-8 RC Time Constants

Explain how to determine the amount of time needed to charge a capacitor that is connected in a circuit with a resistor. Go through the formula and make sure students can use the formula correctly by giving them a couple of practice problems.

19-9 Applications for Capacitors

Provide actual examples or pictures to show students real life applications. Give a general summary of the various everyday uses of capacitors.

19-10 Nonpolarized Capacitors

Describe what a nonpolarized capacitor is and what it consists of. Discuss its uses and advantages over polarized capacitors.

Explain how oil-filled capacitors are marked, what this indicates, and how students should use this information when connecting the capacitor. Make sure students understand why the way in which the capacitor is connected can either lead to or prevent damage to the item the capacitor is connected to.

19-11 Polarized Capacitors

Distinguish between these capacitors and nonpolarized capacitors in their uses and limitations of use. Make sure students realize, however, the difference in capacitance available in a smaller polarized capacitor.

Discuss both the wet and dry types of electrolytic capacitors, and how they are used.

AC Electrolytic Capacitors

Explain the process of shorting out and reforming with this type of wet electrolytic capacitor. Describe when, where, and why this type of capacitor would be needed.

Dry Type Electrolytic Capacitors

Distinguish it from the wet type and describe the gauze make-up of this type of capacitor. Ask students why this internal make-up might be advantageous in a capacitor.

19-12 Variable Capacitors

Compare these capacitors to those with dielectric constants. Explain how the movable plates are used in conjunction with the stationary plates to change the capacitance value.

19-13 Capacitor Markings

Again, show some capacitors and resistors together. It is important for students to feel comfortable distinguishing between the two. Refer back to Figure 5-9 of the text to compare the different color codes of resistors and capacitors.

Discuss the various ways that capacitors are marked, and why these markings vary. Practice on the sample capacitors to see if students can identify the capacitor's voltage, tolerance, etc.

Review the EIA standard and the JAN standard with students. Make sure students understand the difference, and know to look for the identifying white dot that indicates the EIA standard being used.

19-14 Temperature Coefficients

Show students where to find this marking on a capacitor. Make sure they know what positive and negative temperature coefficient markings mean.

19-15 Ceramic Capacitors

If you have samples of these, pass them around and compare them to the other samples. Point out the wide color band, indicating the temperature coefficient.

19-16 Dipped Tantalum Capacitors

Provide samples , if possible, to help identify the different shape of this capacitor. Point out how to use the color bands and dots to determine the various values.

19-17 Film Capacitors

Discuss how these capacitors are different, and how their markings are interpreted.

19-18 Testing Capacitors

Discuss the various ways of testing capacitors. Explain how and why some capacitors need to be tested for both capacitance value and the strength of the dielectric.

Review a few testing procedures. Remind students that some capacitors also need to be tested for leakage. Explain how the dielectric strength test is done, and why it is necessary.

Unit Round Up

Review the key terms and go over the summary in class. Have students do the review at home, and then go over it together during your next class. If you have the materials, this would be a great time for the students to gain some hands on experience. If you do this, warn students, again, about the potential energy stored in a capacitor, and why it is not wise (or funny) to give a charged capacitor to anyone.

Be sure to have students fill in the RC time constants chart at the end of the review section. You may want to reproduce this on the board, and have students come up, one by one, and show how they solved for the missing values.

UNIT 20
CAPACITANCE IN
ALTERNATING CURRENT CIRCUITS

OUTLINE

KEY TERMS

Appear to flow
Capacitive reactance (X_C)
Frequency

Out of phase
Quality (Q)
Reactive power (VARs)

Voltage rating

Anticipatory Set

This unit puts to use the basic information learned in Unit 19. Having a variety of capacitors on hand will add to your explanations.

Go over the key terms and unit objectives with students. Assure them that they are simply taking what they have just learned and putting that knowledge to work. With this in mind, make sure there are no questions, and no confusion left over from the previous unit.

20-1 Connecting the Capacitor into an AC Circuit

Explain what appears to occur when a capacitor is connected to an AC current. Discuss what is actually taking place, and give some examples of capacitors used in an AC circuit.

20-2 Capacitive Reactance

Explain how impressed voltage develops as the electrostatic charge builds and opposes the applied voltage, limiting current flow. Compare this to inductive reactance. Spend a few minutes reviewing inductance. Go over the formula used to solve for capacitive reactance (X_C).

20-3 Computing Capacitance

Go over this formula, making sure students are comfortable with it.

20-4 Voltage and Current
Relationships in a Pure Capacitive Circuit

Discuss the concept of the capacitor charging and discharging at the same time, and how this process results in the current leading the applied voltage by 90°.

Discuss the rotation factor involved with the alternating charges that are taking place, flowing back and forth from positive to negative over and over again.

20-5 Power in a Pure Capacitive Circuit

Discuss the storing of power in the capacitor, and how that power is returned to the circuit when the capacitor discharges. Make sure students realize that in a pure capacitive circuit, there is no true power.

Discuss the type of power existing in a pure capacitive circuit, which is reactive power. Go over the distinction between inductive VARs and capacitive VARs.

20-6 Quality of a Capacitor

Discuss why the quality of a capacitor is usually high, and explain what that means. Go over the formulas that can be used to solve for the quality level.

20-7 Capacitor Voltage Rating

Make sure students realize that you are actually talking about the dielectric when it comes to voltage ratings of capacitors. As you did in Unit 19, warn students of the consequences of exceeding the voltage rating of a capacitor.

Discuss the various ways that voltage ratings can be marked on capacitors. Go through the formula and ensure that students understand how to determine if a voltage rating will be exceeded. The ability to do this with all types of capacitors is crucial. Spend as much time as is necessary to make sure that students can solve for the PEAK voltage rating, and identify when it is exceeded.

20-8 Effects of Frequency in a Capacitive Circuit

Define frequency and explain its effect on capacitive reactance. Discuss how capacitive reactance is inversely proportional to frequency, and what that means to the charge and discharge rate of the capacitor, as well as to the opposition to current flow.

20-9 Series Capacitors

The best way to explain this is to work the problem given in the text. As you work through the problem, preferably by having students come to the board, show comparisons of other solving techniques. Take a minute to look over the Pure Capacitive Circuits Formula in Appendix D of the text.

As you go through this solution, check for understanding in several areas along the way. If students are unable to work through this problem, keep doing problems like this one until they can work through them comfortably.

20-10 Parallel Capacitors

This is another learn-as-you-do section. Using the board, have students come up and solve for the various values. Check for confusion so that you can identify exactly what step throws someone off. Refer students back to Appendix D as it becomes necessary.

You may need to review the process of rounding off numbers.

Unit Round Up

Go back over the key terms and make sure students feel comfortable with their level of workable knowledge of the objectives covered in this unit. Reteach any section that still proves to be confusing. Go through the summary together. Have students do the review questions and fill in the capacitive circuits chart on their own. Go over these together, having students show their work on the board.

UNIT 21
RESISTIVE-CAPACITIVE SERIES CIRCUITS

OUTLINE

KEY TERMS

Angle theta ($\angle\emptyset$)

Apparent power (VA)

Capacitance (C)

Power factor (PF)

Reactive power (VARs)

Total current flow (I)

Total impedance (Z)

Total voltage (E_T)

True power (P)

Voltage drop across the capacitor (E_C)

Voltage drop across the resistor (E_R)

Anticipatory Set

For this unit you may want to do a short review on several of the formulas used to solve various circuit values. You may also want to take a minute to refresh students' memories concerning resistive-inductive series circuits from Unit 17. It will also be very beneficial to have the measuring devices used in this unit on hand for some hands-on experience.

Remind students that what they are learning is simply a combination of skills and applications that they have already learned. Building student confidence always lends itself to a better learning experience.

21-1 Resistive-Capacitive Series Circuits

Compare resistive-capacitive series circuits to resistive-inductive series circuits. Discuss the varying levels at which the voltage and current will be out of phase with each other, and how to determine the exact amount. Go over the list of symbols that students will need to know when working with these circuits, and have students copy them into their notes.

21-2 Impedance

Discuss how total resistance is determined using the combination of resistance and capacitive reactance. Do a few problems illustrating this formula.

21-3 Total Current

Go through this formula, making sure students understand how and when to use it.

21-4 Voltage Drop across the Resistor

Remind students of the rule that the current in a series circuit is the same at any point in the circuit. Work a few problems using the formula for solving for the voltage drops across the resistor.

21-5 True Power

Remind students that both the current and the voltage must be in phase with each other for true power to exist. Go over the formula for solving for true power.

21-6 Capacitance

Do a few problems using the formulas to solve for capacitance.

21-7 Voltage Drop across the Capacitor

Do a few problems using the formula to compute the voltage drop across the capacitor.

21-8 Total Voltage

Review vector addition and do a few practice problems solving for the total voltage.

21-9 Reactive Power

Explain the difference in solving for reactive values with watts versus with voltage, and do some practice problems.

21-10 Apparent Power

Do some practice problems using both the formula in the text and vector addition. Make sure students understand when to use one solution technique instead of the other.

21-11 Power Factor

Explain what the power factor is and do some practice problems using the formula.

21-12 Angle Theta

Review cosine and do a few problems using the formula to solve for angle theta. Review the example in the text, step by step, solving for each value along the way. This should be done on the board, or on the overhead, so that everyone can be involved.

A good way to check for understanding is to ask a student to explain, in his or her own words, how he or she arrived at a particular solution, such as apparent power, total circuit current, power factor, or the cosine of angle theta.

Unit Round Up

Discuss the summary together and have students do the review questions and the practice problems for homework. Have students work the practice problems on the board during your next class session. While students are working these problems on the board, check that they have a clear understanding of what they are doing, and why they are doing it.

UNIT 22
RESISTIVE-CAPACITIVE PARALLEL CIRCUITS

OUTLINE

22-1 Operation of RC Parallel Circuits
22-2 Computing Circuit Values

KEY TERMS

Angle theta ($\angle\emptyset$)
Apparent power (VA)
Circuit impedance (Z)
Current flow through the capacitor (I_C)
Current flow through the resistor (I_R)

Phase angle shift
Power factor (PF)
Reactive power (VARs)
Total circuit current (I_T)
True power (P)

Anticipatory Set

This unit will be similar in concept to the last unit. A short review of parallel circuits, cosines, and the tables in Appendix A and Appendix B may be a good way to begin. Go over the unit objectives and key terms to prepare students for the unit. Remind students, again, that this unit combines many things that they already know.

Begin by explaining that they will be learning the effect of the current flow on other circuit quantities. Any test instruments that you can provide for hands-on learning will be invaluable as a teaching tool. This unit is also a good unit to use to have students explain to you why these various formulas work, and how they know which formula to use to obtain certain values.

22-1 Operation of RC Parallel Circuits

Make sure that students are aware of the difference in the voltage across all devices and the current flow through the capacitor. Explain phase angle shift and how it will be determined.

22-2 Computing Circuit Values

Draw Figure 22-2 on the board or overhead so that you can walk the class through each step of the solution process. Have students enter the symbols for each quantity into their notes for easy reference.

Resistive Current

This should be a familiar formula, so simply do this together on the board or overhead.

True Power

Discuss how using any of the values associated with the pure resistive part of the circuit can result in the amount of true power. Show some examples, and then go through the formula given in the text. Double check for an understanding of what true power really is.

Capacitive Current

Go through the formula and make sure students understand what capacitive current is.

Reactive Power

Go through the formula with the class, checking for understanding by calling on various students for explanations.

Capacitance

Go through the formula, checking for understanding by having students explain why this formula works.

Total Current

Remind students that the voltage is the same across all legs of a parallel circuit. Discuss the out of phase aspect of the current flow through the capacitor, and the effect of the voltage lagging that current by 90^0. Go through the formula, checking for understanding.

Impedance

This formula should also be familiar to students, reminding them of the Ohm's law formula. Discuss this formula, and when and how to use vector addition to come up with the same solution.

Apparent power

Go through this formula and check for understanding.

Power Factor

Explain what the power factor is, and go through the formula.

Angle Theta

Explain how angle theta is formed between the apparent and true power, and why the cosine of angle theta is equal to the power factor. Draw the next circuit on the board or overhead, and have students copy the symbols for the various values into their notes. Go through the formula, as you did with the last section.

True Power

Walk students through both formulas, and check for understanding.

Reactive Power

Refer to Figure 22-9 and draw this one on the board or overhead. Be sure that students understand how this right triangle is formed. Remind students that they learned how to use the Pythagorean theorem in Unit 14, and take them through the process of solving for reactive power.

Capacitive Reactance

Walk students through this formula, checking for understanding.

Voltage Drop across the Capacitor

Make sure students are clear on what they are solving for, and why this formula works to obtain that solution.

E_R and E_T

Remind students that the voltage is the same across all the branches of the parallel circuit, and explain why E_T and E_R are, therefore, the same.

I_C

Work through this formula, asking students to explain why this formula works.

I_R

As with each of the previous formulas, ensure that students understand why this particular formula is used to obtain this quantity.

Resistance

Note the similarity to Ohm's law, and work through this on the board or overhead.

Total Current

Explain how total current can be solved for by using Ohm's law. Then, take students through the solution process, using vector addition.

Impedance

Go step by step through the formula and, as always, make sure students understand why this is the formula to use for solving for the impedance value.

Unit Round Up

Go through the summary in class, and have students explain the key terms covered in this unit. Have students answer the review questions and work the practice problems on their own. In the next class session, ask students to work these out on the board, explaining what they did each step of the way. Double check for any confusion.

Make sure each student can solve for the values in a circuit of this type. Continue solving practice circuits until all students are comfortable with knowing when to use a particular formula, and how to use those formulas.

SECTION 8
Alternating Current Circuits Containing Resistance-Inductance-Capacitance

UNIT 23
RESISTIVE-INDUCTIVE-CAPACITIVE SERIES CIRCUITS

OUTLINE

23-1   RLC Series Circuits
23-2   Series Resonant Circuits

KEY TERMS

Bandwidth
Lagging power factor

Leading power factor
Resonance

Anticipatory Set

Review formulas that have been used in Units 21 and 22, as they will be applied in this unit. Review alternating current circuits, resistance, inductance, and capacitance.

Ask students to take turns explaining each of the terms they should now be familiar with. These terms should include voltage drop, resistance, inductance, capacitance, impedance, true power, reactive power, angle theta, cosine, and power factor. When you have established that students have a good working knowledge of these terms, alert them that you will, again, have them explaining things to you, throughout this unit. As in Unit 22, you want to be sure students understand, on their own, which formula works in specific situations, and why.

23-1 RLC Series Circuits

Explain why the current is the same in an alternating current when the resistance, inductance, and capacitance are connected in series. Then, discuss why the voltages dropped across the elements are out of phase with each other. Refer to Figure 23-1 as you go through these explanations. Show the relationships between the voltage and the current as it drops across the resistance, the inductor, and the capacitor. Explain how the applied voltage will be determined, and why it lags behind or leads the circuit current. Explain the meaning of the terms lagging power factor and leading power factor.

Discuss the effect of the inductive reactance being $180°$ out of phase with the capacitive reactance, resulting in a high amount of current flow. Draw a sample problem on the board or overhead and refer to it as you proceed through the solution process.

Total Impedance

Explain what the total impedance is derived from, and why vector addition must be used to solve for it. Go over the process together and make sure students understand that arriving at a negative answer inside the parenthesis won't matter because that amount will be squared.

Current

This formula should be very familiar to students, but check for understanding to be sure.

Resistive Voltage Drop

This is another familiar formula, but check for simple errors in multiplication.

Watts

Work with the formula in the text, and then use another pure resistive value to show how any pure resistive value within the circuit can be used.

Inductance

Go through the formula, ensuring that students know what each of the symbols represents.

Voltage Drop across the Inductor

This is a familiar formula, and students should be able to work through it easily. Be sure that students realize that the voltage drop across the inductor is greater than the applied voltage. Have students explain why this is true.

Inductance VARs

Make sure students understand what they are solving for here, and that they know when to use this formula.

Capacitance

This formula offers a good opportunity to make sure that students can identify what each symbol represents. Also, be sure they understand why to use the specific elements they are using to obtain the capacitance value. In other words, why would vector addition, or some other formula, not work here?

Voltage Drop across the Capacitor

This formula should be familiar, so just walk students through and have them explain why this formula works.

Capacitive VARs

Compare this formula to the one for computing inductance VARs, and have students explain the differences.

Apparent Power

Go through both formulas and have students explain why one formula is better in some situations, and why the other formula is better in others. Also, have students explain the difference between apparent and true power, and why true power can, therefore, be used to compute apparent power. Refer to Figure 23-5 during this discussion.

Power Factor

First, have someone explain what the power factor is. Then, have one student explain why this is the correct formula to use, and another student explain why to multiply by 100. Have yet another student explain the importance of obtaining the percentage.

Angle Theta

Make sure students know what a cosine is and how to use the tables in the Appendix. Then, go through each step of the solution process. Draw the circuit on the board or overhead, so that you can fill in the values as the class solves for them.

23-2 Series Resonant Circuits

Begin with a discussion of what resonance is and how it is achieved. Explain why the increase in frequency results in an increase in inductive reactance but a decrease in capacitive reactance. Discuss why resonant circuits are used to increase current. Compare this method for achieving large increases in current to other methods, and be sure students understand the concept of using resonance for this purpose. Use actual voltmeters and a circuit to illustrate, if posible

Review the formula, ensuring that students are comfortable with it. Discuss the process of recognizing what effect an increase in frequency will have on inductive reactance and capacitive reactance.

Bandwidth

Explain what bandwidth is and how this frequency measurement is used to determine the rate of increase or decrease in current. Refer to Figures 23-13 and 23-14, having students explain what is taking place in each example.

Determining Resonant Values of L and C

Discuss when it will be necessary to determine inductance or capacitance values needed in order to resonate with existing values for inductance or capacitance. Go carefully over the formulas, and do a few practice problems, checking for understanding.

Unit Round Up

Go through the summary together. It is vitally important that students can explain each of the terms covered in this unit. It is also important to review the various fomulas and to make sure students know when to use each one, and why. Have students do the review questions and practice problems on their own, then on the board, explaining what they did, and why they used the solution process that they used. If there is any confusion, go back over the material that is not clearly understood.

UNIT 24
RESISTIVE-INDUCTIVE-CAPACITIVE PARALLEL CIRCUITS

OUTLINE

24-1   RLC Parallel Circuits
24-2   Parallel Resonant Circuits

KEY TERMS

Bandwidth   Tank circuits
Power factor correction   Unity

Anticipatory Set

Go over all terms and an overview of the unit, letting students know that this unit will be adding to what they have already learned in Units 22 and 23. As always, having all measuring devices used in this unit will be invaluable. Additionally, be sure to draw all sample circuits from the text on the board or overhead. As you walk students through each formula, utilize these drawings to fill in quantity amounts. A short review of Q and VARs from Unit 16 might be useful for students.

24-1 RLC Parallel Circuits

Discuss why the voltage dropped across each element remains the same, while the currents flowing through each branch are out of phase with each other. Emphasize that this is a primary characteristic of an AC circuit that contains elements of resistance, inductance, and capacitance, connected in parallel. Then, point out, using Figure 24-1, just how the various current flows lag behind, lead, or are equal to other elements of the circuit. Have students explain each of these in phase and out of phase combinations.

Compare the inductive and capacitive VARs in a parallel circuit to those in a series circuit, and make sure students can distinguish between the two.

Copy the RLC parallel circuit in Figure 24-2 on the board or overhead, and refer to it as you proceed through the solution process. Ask students to close their books, and then give them each formula as they need it.

Don't fill in any quantities for them, and see if they can use what they learned in Units 22 and 23 to obtain the quantities they need to fill in the circuit. Make sure, first, that they have copied down the basic information and all of the symbols needed to do the formulas. If this tactic doesn't appear to be working, then go back to advancing, formula by formula, through the book.

Impedance

Review this, and have students compare this process with the formula they used to solve for total impedance in an alternating current RLC series circuit.

Resistive Current

Point out that this is a basic Ohm's law-type of formula, and should be very familiar to them.

True Power

This is another familiar formula and should require only a brief mention.

Inductive Current

Briefly go through the formula and then have students explain the difference between this and solving for current in an alternating current RLC series circuit. Have students explain why an alternating current RLC series circuit has straight current to solve for, but the parallel circuit has resistive and inductive current to solve for.

Inductive VARs

Go through the formula, and have students compare this formula to the one for inductance VARs in alternating current RLC series circuits.

Inductance

Review this formula with the students and have them explain why the same formula can be used for both series and parallel circuits of this type.

Capacitive Current

Go through the formula and have students explain why it is so different from the current formula in Unit 23.

Capacitance

As with inductance, work the formula and have students explain why the same formula works for both series and parallel circuits of this type.

Capacitive VARs

Work through this formula and compare it to the one used in Unit 23.

Total Circuit Current

Explain that the inductive current and the capacitive current cancel each other out. Then take students through the formula, referring to Figure 24-4. Work the alternative method for computing total current, and discuss when one formula would be better to use than the other. In other words, if the second formula works and is so similar to an Ohm's law formula, why would a student need to use vector addition?

Apparent Power

Work through both formulas and compare them to those used in Unit 23 for a series circuit of this type.

Power Factor

Review the formula and have students explain why the same formula works for both series and parallel circuits of this type.

Angle Theta

Go through this solution, noting it is the same for both series and parallel circuits of this type.

24-2 Parallel Resonant Circuits

Begin by comparing series resonant circuits to parallel resonant circuits, pointing out the cancelling out effect that inductive and capacitive current have on line current. Using Figures 24-7 and 24-8, explain what tank circuits are.

Explain the role that the quality (Q) of the circuit components has in both total circuit current and total circuit impedance. Focus on the inductor in this explanation. Ask students to explain why the inductor is such an important factor.

Work with students through the process of finding the total circuit current at resonance. Discuss how to do this using the true power caused by the resistance of the coil, as well as Ohm's law. Refer to Figure 24-9 as you work through this process.

Bandwidth

Compare the process of determining bandwidth in both series and parallel circuits of this type. Be sure students notice that circuits with a high Q will have a narrow bandwidth, and circuits with a low Q will have a wide bandwidth. Have students explain why this is.

Induction Heating

Explain what induction heating is and how this type of heat is obtained using a tank circuit. Also, discuss why this would be a preferable way of heating metal. Discuss how this tempered heat can be controlled by adding or subtracting capacitance in the tank. Ask students if they can think of other uses for this type of heat source.

Power Factor Correction

Explain what power factor correction is, why it is used, and where it would be used. Discuss the difference between the heat energy and mechanical energy produced in the circuit. As part of the discussion, make sure students notice that only the resistive value of the motor changes as load is added or removed.

Explain how to determine the existing motor power factor (apparent power), and then the power factor, using the formulas given. Now that the power factor can be found, explain why the correction is needed, and how students will be able to distinguish when a power factor correction is necessary. Walk students through the formula for determining what part of the circuit is reactive.

Explain what unity means in relation to power factor correction, and how it can be achieved. Make sure students understand why achieving 100% is not practical, and why, therefore, a 95% power factor is the aim. Review the formula to determine the apparent power needed to reach 95%, and then work through the formula to determine the amount of inductive VARs needed to obtain the correct amount of apparent power. Compare this formula with the reactive power formula from Unit 22. Have students explain the inherent differences in the two formulas and why each is used as it is. This is a good way to check for understanding. Go through each step of this process, all the way to determining how to achieve a 95% correction level, checking for understanding as you go.

Unit Round Up

Go over the summary together and assign the review questions and practice problems for class work or homework. Go over these assignments in class, asking students to explain how they got their answers, and to show their work on the board.

UNIT 25
FILTERS

OUTLINE

KEY TERMS

Bandwidth
Filter circuit
Notch filters
One tenth
Pass

Peak
PI (π) filter
Q of the interior
Reject
Resonant

Tank circuit
Ten times
Trimmer Capacitor
Trimmer Inductor

Anticipatory Set

This unit will combine students' knowledge of inductor and capacitor use in achieving resonance, and the frequencies that can result. Your examples may be verbal, audio, and visual, including a television, a radio, and other examples. Compare for example, an AM/FM radio and a ham radio.

Go over the unit objectives and key terms. Remind students that they are simply taking what they learned in Units 22-24 one step further. Briefly discuss that resonant circuits, which students are already familiar with, are one type of filter. Refer to Figure 25-1 and explain the difference between a trimmer inductor and a trimmer capacitor.

25-1 Broad-Band Tuning

Explain the difference between regular inductors and capacitors and variable inductors and capacitors. Discuss why switching over to a variable inductor or capacitor allows for a broader band. Compare this to what students learned about variable resistors in Unit 5. Ask students to explain why variable capacitors are used more often than variable inductors.

25-2 Low-Pass Filters

Explain what a low-pass filter is and how it functions. Discuss the differences between hooking up an inductor in series with a load resistor and hooking up a capacitor in parallel with the load resistor. Explain how these two different systems work, and then discuss the low-pass filter that uses both a capacitor and an inductor.

25-3 High-Pass Filters

Emphasize that the hook-up of the inductor and capacitor is reversed when creating a high-pass filter. That is, the capacitor is now connected in series with a load resistor, and the inductor is connected in parallel. Discuss how these two connection patterns result in filtering processes.

25-4 Bandpass Filters

Explain how a bandpass filter is created, empasizing why the bandwidth is determined by the Q of the inductor. Also, discuss the use of the tank circuit connected in parallel with the load resistance. Have students explain this whole process back to you, as a way of checking for understanding.

25-5 Band-Rejection (Notch) Filters

Discuss both methods to create band-rejection filters. Discuss under what circumstances these filters would be preferred over bandwidth filters.

25-6 T Filters

Explain what a T filter pattern looks like, and describe the components that are normally found in a T filter.

25-7 PI Type Filters

Explain why filters connected this way are called PI filters, and compare them to T filters.

25-8 Crossover Networks

Explain what a crossover network is and how it is used. Point out that a crossover network takes the best filter patterns for each desired frequency and puts them together for one overall effect. Have students share experiences they may have had with speakers blowing out, and why this may have occurred. Ask students if an incorrectly wired filter could cause speakers to perform poorly or to blow out.

Unit Round Up

Go over the summary together and have students do the review questions and practice problems on their own. Discuss these during the next class session, and check for understanding by having students explain how they came up with the answers they have.

Section 9
Three-Phase Power

UNIT 26
THREE-PHASE CIRCUITS

OUTLINE

KEY TERMS

Correcting the power factor

Delta connection

Line current

Line voltage

1.732

Phase current

Phase voltage

Star connection

Three-phase watts

Three-phase VARs

Wye connection

Anticipatory Set

Go over the unit objectives and key terms to familiarize students with the material. As usual, any visual examples that you can have in the classroom will be very helpful. A brief review of Units 4 and 13 will be beneficial for those who have forgotten some of the principles of polarization.

26-1 Three-Phase Circuits

Discuss the three main reasons that three-phase power is better than single-phase power. Refer back to Unit 24 and have students try to imagine what they would have to do differently to determine the power factor correction amount using a single-phase power circuit rather than a three-phase circuit. Why might there be a difference? How could this fluctuation difference in a single-phase circuit alter other quantity measures in the circuit?

Explain how the three-phase voltage is produced when the three magnetic fields cut through the coils.

26-2 Wye Connections

Draw a wye connectionon the board or overhead as you begin your explanation. Explain what a wye connection is and what phase and line voltages are. Then show students how to use 1.732 to determine the amount of line voltage or phase voltage. Have students make a note to remember that in a wye-connected system, the line voltage will be higher than the phase voltage, but that both line and phase current will be the same.

Voltage Relationships in a Wye Connection

Compare the differences between the single-phase system and the three-phase system, explaining why the line voltage is 208V in a three-phase system, instead of 240V.

26-3 Delta Connections

Describe what a delta connection is, making sure students realize that three separate inductive loads are used. Also point out that although the line and phase voltage are the same, the line current is higher than the phase current. Go over the formulas for determining the current, and review the vector addition to determine the sum of the currents.

26-4 Three-Phase Power

Discuss both formulas and emphasize the major difference between the two. Make sure students can determine when to use the square root of 3 as the multiplier, and when to use the whole number 3. If you have a voltmeter and clamp-on ammeter available, and a three-phase circuit that you can hook on to, demonstrate the use of the first formula.

26-5 Watts and VARs

Go through each formula, distinguishing which formula to use for watts and which one to use for VARs, and do a few of these on the board.

26-6 Three-Phase Circuit Calculations

Have students copy the symbols for these calculations into their notes. Then, guide students through the solution process, step by step, checking for understanding as you go. Be sure that students are aware of the differences in the solution process when the connections are set up differently. For example, distinguish the solution process of a wye connection from that of a delta connection. As you proceed through each example, be sure to have students explain back to you what is taking place, and why the formulas being used are the correct ones needed for the particular solution being sought.

26-7 Load 3 Calculations

Define what a load 3 calculation is, and what it is used for. Work through the formulas, and make sure students understand that they are dealing with a wye connection.

26-8 Load 2 Calculations

Define what a load 2 calculation is, and have students note that, unlike load 3 calculations in which they were dealing with capacitors, they will now be dealing with inductors. Additionally, they are now solving using a delta connection.

26-9 Load 1 Calculations

Point out the differences between a load 1 calculation versus a load 2 calculation. Also, make sure students note that they are now working with resistors in a wye connection. Compare this set-up to the load 3 calculation, in which they were working with capacitors in a wye connection.

26-10 Alternator Calculations

Explain the role of the alternator, and take students through the formulas. Remind students that when they are solving for the circuit power factor, they will be rounding off, and writing their answers as percentages.

26-11 Power Factor Correction

Review the process for power factor correction in a single-phase circuit. Point out that a clamp-on ammeter and a voltmeter are used to solve for quantities in the next example. This should alert students that the multiplier will be the square root of 3. Take students through the solution process, checking for understanding along the way.

Unit Round Up

Go through the summary together. Review the terms, making sure students can explain the difference between a wye and a delta connection. Also, be sure that students can explain the differences in formulas for solving load 1, 2, and 3 calculations. Have students do the review questions and practice problems on their own. Go over these during your next class session, having students show their work on the board while explaining what they are doing, and why.

Section 10
Transformers

UNIT 27
SINGLE-PHASE TRANSFORMERS

OUTLINE

KEY TERMS

Autotransformers

Control transformer

Distribution transformer

Excitation current

Flux leakage

Inrush current

Isolation transformers

Laminated

Neutral conductor

Primary winding

Secondary winding

Step-down transformer

Step-up transformer

Tape wound core

Toroid core

Transformer

Turns ratio

Volts-per-turn ratio

Anticipatory Set

Go over the unit objectives and all key terms. Students will be dealing with polarity identification so, if you feel it is needed, briefly review Units 4 and 13, and emphasize the properties of magnetism and magnetic fields. Also, review the effect of coil windings on inductance. Even though you did this in Unit 26, it is always good to begin a new unit by allowing your students to become aware of how much they already know.

Have a variety of transformers on hand to show students, and a voltmeter for demonstration purposes. Even though you just covered single-phase circuits, do a brief review of those formulas and elements to further build student confidence.

27-1 Single-Phase Transformers

Give an overview of transformers and their uses, listing the three classifications of transformers. Explain how the values of a transformer are proportional to the turns of wire, per winding, of the primary and secondary windings.

Transformer Formulas

Have students copy the symbols that will be used in the upcoming formulas into their notes for easy reference. Distinguish between the primary winding and the secondary winding. Refer to Figure 27-2 while making these distinctions.

27-2 Isolation Transformers

Define what an isolation transformer is, and what it is used for. Make sure students understand the difference between windings magnetically coupled versus electrically coupled. Explain how this magnetic coupling results in a filtering effect. Discuss the voltage spike that can occur, and why this is reduced in an isolation transformer. Explain the construction of the isolation transformer, and the benefit gained by the stacked laminations.

Basic Operating Principles

Discuss how isolation transformers work following the expansion and collapse of the magnetic field. Define self-induction.

Excitation Current

Explain what excitation current is and that it occurs regardless of the type or size of the transformer.

Mutual Induction

Define mutual induction and discuss the ratio of the number of turns of wire in the secondary to those in the primary, and how this results in a volts-per-turn ratio.

Transformer Calculations

Take students step by step through the formulas, ensuring that they understand why each formula works to give them the needed value. Explain what a step-down transformer is and when a transformer of this type is used.

Go over the rule that power in must equal power out (you can compare this to a computer saying, "Garbage in-garbage out."). Help students understand that this is a good way to determine if their calculations are correct. If the primary volt amps are not equal, then they know that they have miscalculated something.

Define a step-up transformer, and be sure students can distinguish this type of transformer from a step-down transformer.

Calculating Isolation Transformer Value Using the Turns Ratio

Explain how to determine the turns ratio by dividing the higher voltage by the lower voltage. Show students how to use Ohm's law to determine the amount of secondary current. Then show students how to determine the primary current, and to check their calculations by making sure that the primary and secondary volt amps quantities are equal.

Multiple-Tapped Windings

Explain what multiple-tapped windings are, and display this with your sample transformers. Also, tell students where multiple-tapped windings are used. Explain how the turns ratio is maintained in spite of the taps. Do the same thing for multiple taps on the secondary. Also discuss the advantage of electrically isolating multiple secondary windings from each other. Explain where a transformer with multiple secondary windings is used.

Computing Values for Isolation Transformers with Multiple Secondaries

Explain that to do the computations for a transformer that has a secondary with multiple windings, treat each winding as a separate transformer. Take students through the steps to accomplish this task.

Distribution Transformers

Explain what a distribution transformer is and the role it plays in electricity going into businesses and residences. Ask if students have ever been the victims of a transformer on a pole or in their neighborhoods going out with a bang. The obvious result is a loss of power to the houses tapped into that transformer.

Describe how the neutral conductor works to allow both 120V and 240V current to be available at the same time. Refer to Figures 27-17 and 27-18 to help illustrate this. Ask students what could happen if loads requiring 240V are connected on only one side of the neutral conductor. Then, ask students what would happen if a load requiring 120V were connected directly across the lines of the secondary. Since these students will ultimately be making these connections in the future, it is not only important for them to know what to do, but also to know what *not* to do.

Control Transformers

Explain how a control transformer works and where this type of transformer is used. Refer to Figures 27-26 and 27-27 while you describe the way the primary windings are hooked up. Discuss the turns ratio and how it is affected by the series-connecting of the primary windings.

Transformer Core Types

Discuss the laminated core and why it is preferred over other types. Have students explain why the reduction of eddy currents is important. Explain the role the power rating of the transformer plays in the amount of core material in the transformer. Describe the various core type transformers and the uses and benefits of those various types. Distinguish between each one also by explaining why one type is better than another, depending on the type of load that will be applied to the transformer.

Describe the tape wound core or toroid core transformer, and the benefit gained by the core being one continuous length of metal.

Transformer Inrush Current

Make the distinction between the operating characteristics of reactors and transformers, as you explain what inrush current is.

Magnetic Domains

Explain what magnetic domains are and how their molecular shapes are charged by outside magnetic forces. Refer to Figures 27-34 and 27-35 to illustrate this reaction. Also, discuss how this reaction is reversed back to normal structure in some domains when the outside magnetic force is removed. Explain why this type of domain is used in reactors and chokes, or have students try to explain this to you.

Discuss the inrush factor of transformers that "remember" where they were last set, and how each factor limits the inrush current.

27-3 Autotransformers

Describe autotransformers and discuss how they are used. Compare autotransformers to isolation transformers. Explain how the turns ratio can become a variable ratio, and why that is beneficial. Also, explain why an autotransformer could present a problem due to the lack of line isolation between the incoming power and the load.

27-4 Transformer Polarities

Explain what transformer polarity is and how the voltage produced across a winding determines this polarity.

Polarity Markings on Schematics

Using Figure 27-45, point out and explain how to locate the polarity of transformer windings on a schematic. Provide copies of some schematics for students to practice with.

Additive and Subtractive Polarities

Explain what additive and subtractive polarities are, and familiarize students with the terms buck and boost. Describe how a buck or boost is tested for, and show this in the classroom using your voltmeter and transformers.

Using Arrows to Place Dots

Explain the use of arrows to indicate a buck or boost polarity, and point this out on sample schematics handed out to students.

27-5 Voltage and Current Relationships in a Transformer

Explain the relationship of the primary current to the voltage in the secondary. Be sure students understand how this is affected during the times that the primary current is changing the most, and the overall effect this has on the secondary voltage lagging the primary current.

Adding Load to the Secondary

Describe what happens when a load is connected to the secondary. Point out that this load on the secondary begins the flow of current, and that the increase in current in the secondary will result in a proportional increase in current in the primary. Connect a wattmeter to the primary transformer you have set up in this way to illustrate the true power increasing as load is added to the secondary.

27-6 Testing the Transformer

Explain how to use an ohmmeter to test for grounds, shorts, and opens. Demonstrate this with an ohmmeter and transformer in the classroom. Do the same thing using a megohmeter. Allow students to have some hands-on experience using these meters to test the transformers you have available. Make sure students understand what they're testing for with each device.

27-7 Transformer Ratings

Describe the information students can find on the nameplate of a transformer. Allow students to look at the nameplates on the transformers you have, so they can become familiar with looking for and identifying this information.

27-8 Determining Maximum Current

Point out where the information needed to find the maximum current can be found on the nameplate. Then, explain how to use the formulas to obtain the values for secondary voltage and primary current.

27-9 Transformer Impedance

Explain how transformer impedance is determined, and how each factor can affect the level of impedance. Point out how impedance is expressed (%Z or %IZ), and how to measure it, using the short circuit connection and the variable voltage source connection to the high voltage winding. Go through the process of calculating the transformer impedance. Walk students through the examples and formulas given, checking carefully for understanding.

Unit Round Up

Go over the summary together in class. Take some time with this, having students give the explanations or elaboration on the statements made in the summary. Have students do the review questions and practice problems on their own. During your next class session, have students answer the questions and give all explanations from the review. Then, ask students to work the practice problems on the board, as they explain what they are doing and why they are doing it. If you detect confusion during this time, reteach whatever areas seem unclear.

UNIT 28
THREE-PHASE TRANSFORMERS

OUTLINE

KEY TERMS

Closing a delta
Delta-wye
Dielectric oil
High leg

Neutral conductor
One-line diagram
Open delta
Orange wire

Single-phase loads
Tagging
Three-phase bank
Wye-delta

Anticipatory Set

Go over the unit objectives and key terms. Do a quick review of the wye and delta hook-ups as preparation for the three-phase connections in Unit 28. As always, have the trasformers on hand to illustrate the explanations given in the unit. Allow students to make wye and delta connections with the transformers for practice to cement this process in their minds.

28-1 Three-Phase Transformers

Explain what a three-phase transformer is and discuss the use of the dielectric oil.

Three-Phase Transformer Connections

Discuss the two types of connections. Emphasize that the ordering of the name of the transformer will dictate the wiring pattern.

Connecting Single-Phase Transformers into a Three-Phase Bank

Explain how to form a three-phase bank using three single-phase transformers. Using Figures 28-5, 28-6, and 28-7, help students identify the windings and connections needed to create the three-phase transformer. Discuss how each connection is made, and why this pattern is used. Have students explain this back to you to check for understanding.

Using Figure 28-9, explain and discuss one-line diagrams. Discuss when this schematic might be enountered rather than the previous ones studied.

28-2 Closing a Delta

Emphasize why it is important to check for proper polarity before making final connection and applying power. Have students explain what would happen if the proper polarity connections were not in place.

Explain and demonstrate how to use the voltmeter to check for proper phasing. Make sure students realize that the voltmeter could register some amount of voltage even before the delta is closed.

28-3 Three-Phase Transformer Calculations

Have students copy the symbols needed for the upcoming formulas into their notes. If you think it is needed, briefly review the formulas for transformer calculations and three-phase calculations. Be sure students take note of the rule that only phase values of voltage and current can be used when computing transformer values.

Guide students through the solution process, having them explain to you what is taking place each step of the way. Make sure students can do this for both connection patterns: wye-delta bank and delta-delta.

28-4 Open Delta Connection

Discuss the open delta and closed delta banks, noting how they differ. Also, discuss the output power of each. Give examples of under what circumstance each of these connections would be used, and why one of these would be chosen over a three-phase transformer, even though the power output is lower. Go over the computation process that is used.

28-5 Single-Phase Loads

Explain what a single-phase load is and why three-phase transformers are used for this purpose, even though their use could cause unbalance in the system.

Open Delta Connection Supplying a Single-Phase Load

Referring to Figure 28-17, explain how this open delta connection with one center-tapped transformer works. Make sure students understand the different load limitation ratings, and how to use these capacities to obtain the total power needed.

Voltage Values

Using voltmeters on a set-up like the one in Figure 28-17, demonstrate that the voltage is the same across the three-phase lines. Do the same for the 120V lines.

Have students tell you which lines are supplying the single-phase power. Explain what the high leg line is and how to compute this voltage. Also, point out the use of the orange wire or tagging process used to identify this high leg wire. Have students explain why it is important for this wire to be so easily identifiable.

Load Conditions

Explain the various load conditions available, based on whether the single-phase load or three-phase load is in operation. Be sure that students understand the function of the neutral conductor. Have this set up so that you can connect ammeters, demonstrating this process for students. Also, direct students to note the different functions of the smaller and larger transformers, then discuss how the high leg can be used to supply the single-phase load.

Calculating Neutral Current

Referring back to the role the neutral conductor played in the previous example, explain how to calculate the neutral current.

28-6 Closed Delta with Center Tap

Explain how similar this connection is to the one students have just dealt with, except for the addition of the third transformer. Discuss why it is beneficial to close the delta.

28-7 Closed Delta without Center Tap

Explain how this system is set up, and what the benefits are of this type of connection. Give examples of under what circumstances a closed delta without center tap would be used.

28-8 Delta-Wye Connection with Neutral

Explain what this connection looks like and why it is the most commonly used for single-phase loads. Point out the role of the neutral conductor. Emphasize that if this type of transformer connection is powered by a three-phase wire system, it is important that the primary winding be connected in a delta configuration. Discuss what can happen if this is not the case. Refer to Figures 28-21 and 28-22 during this discussion.

28-9 T-Connected Transformers

Explain the roles of the main transformer and the teaser transformer in supplying three-phase power. Also, referring to Figures 28-23, 28-24, and 28-25, explain how to make the T connection. Compare this connection to the open delta connection.

28-10 Scott Connection

Explain how this connection works and compare it to the T connection. Discuss when this connection is more useful than a T connection.

28-11 Zig-Zag Connection

Explain the use of this connection and how it is created.

28-12 Harmonics

Explain what harmonics are, and how to determine the amount of Hz for each harmonic. Discuss the production of three phase harmonics and be sure students understand what a pulsating load is. Refer to the bridge rectifier, on page 787, and explain the role it plays, and how the filter capacitor is used to smooth the pulsations. Point out what can happen to the AC sine wave.

Harmonic Effects

Discuss the detrimental effects harmonics can have on equipment, and point out symptoms that are tell-tale signs of a harmonic effect. Explain why they need to be able to troubleshoot this type of information as electricians. Differentiate between a positive and negative sequence of rotation. Explain what triplens are and refer to the chart in Figure 28-32. Discuss the problem triplens can cause in a three-phase four-wire system, when they disrupt the normal phasor relationship of the phase currents. Also, talk about how harmonic currents can adversly affect transformers and stand-by generators.

In addition to discussing heating problems in equipment, point out how heat may also trip thermal-magnetic circuit breakers. Also, go over how the different frequencies produced by harmonics can affect electrical panels and buss ducts. Explain why cable wires are run as far from phase conductors as possible, and the high pitched buzzing sound that is emitted from them.

Determining Harmonic Problems on Single Ph-se Systems

Go through each step that an electrician should take to determine if they are dealing with a harmonics problem. Discuss the types of ammeters that will be needed to obtain a correct RMS current value. Explain how to use a regular ammeter and an RMS ammeter to measure the amperes, and determine the ratio to end up with a true RMS value. Make sure students understand how to obtain the ratio, and how to use that ratio to determine whether they have a harmonics problem.

Determining Harmonics Problems on Three-Phase Systems

Explain the similarities in determining harmonic problems in single-phase and three-phase systems. Refer to the chart in Figure 28-38 as an aid in explaining how to read the values and see the problem.

Dealing with Harmonic Problems

Walk students through each step of how to try to correct a harmonics problem, since removing harmonics causing equipment will probably not be feasible.

Determining Transfer Harmonic Derating Factor

Explain what the instantaneous peak phase current is, and how to measure it. Go through the formula used to determine the transformer harmonic derating factor. Show students how to use the kVA capacity of the transformer to determine the maximum load that can be placed on the transformer, and what it should be derated to, in order to function properly.

Unit Round Up

Go over the summary together in class, having students discuss each of the points being summarized. Require students to complete the review questions and practice problems on their own. At your next class session, have students answer the review questions out loud, and work the problems on the board, explaining why they solved the problems the way they did. Make sure students are familiar with all key terms from the unit.

SECTION 11
Direct Current Machines

UNIT 29
DIRECT CURRENT GENERATORS

OUTLINE

KEY TERMS

Armature	Cumulative compound	Pole pieces
Armature reaction	Differential compound	Series field diverter
Brushes	Field excitation current	Series field windings
Commutator	Frogleg wound armatures	Series generator
Compound generators	Generator	Shunt field windings
Compounding	Lap wound armatures	Shunt generators
Countertorque	Left-hand generator rule	Wave wound armatures

Anticipatory Set

Have a variety of generators on hand for student use. Provide a brief review of direct current. Make sure students remember that it flows in only one direction (unidirectional), unlike alternating current which can reverse its direction (bidirectional). Review Ohm's law, reminding students that in a direct current circuit, the current is directly proportional to the voltage and inversely proportional to the resistance. Also review magnetic induction from Unit 13.

Discuss the many uses of DC generators and why, in some cases, they are a better choice than AC generators. Go over the unit outline and key terms that students will need to be familiar with by the end of the unit.

29-1 What is a Generator?

Define generator and discuss magnetic induction with regard to generators. Explain how this induction process is affected with each 90^0 turn of the shaft.

Point out the armature and the fact that the voltage produced in the armature alternates polarity. Have students write down the rule about voltage produced in all rotating armatures, and the use of the commutator.

Referring to Figures 29-8 through 29-16, explain how the commutator works. Also, be sure students gain a clear understanding of what the neutral plane is and why they need to be able to identify this position before setting the brushes. Explain rectified DC voltage and how it is achieved. Also, explain ripple pulsation and how it occurs.

29-2 Armature Windings

Lap Wound Armatures

Have students put into their notes that lap wound armatures are used when low voltage and high current is required. Show some of these armatures while you discuss their construction and use.

Wave Wound Armatures

Make sure students note that this type of armature is used when high voltage and low current are required. Compare the wave wound armatures to the lap wound armatures, as they are exactly opposite in construction and use.

Frogleg Wound Armatures

Explain why these are the most commonly used, and have students compare them to the two previous armatures.

29-3 Brushes

Explain how brushes are used, what they are made of, and how they are usually labeled.

29-4 Pole Pieces

Explain the role of the pole pieces in providing the magnetic field needed to operate the machine. Discuss their construction and have students explain why they are constructed of this material. Mention the magneto and explain how it differs from a regular DC generator.

29-5 Field Windings

Discuss series field and shunt windings, pointing out their differences in use, types of wire used, and resistance. Explain when one would be better to use than the other.

29-6 Series Generators

Explain what a series generator is and how it works. Be sure students understand that the pole pieces need to have some residual magnetism. Discuss the three factors that determine the output voltage. Have students explain the effect of the wound wire on induction, increasing the lines of flux, and increased armature speed on the series generator.

Connecting Load to the Series Generator

Explain the effect of connecting a load to the series generator, emphasizing the continual strengthening of the magnetism of the pole pieces. Have students explain what this strengthening will result in, up to the point of pole saturation.

Give examples of where series generators are used, and why they are used in these instances instead of shunt generators.

29-7 Shunt Generators

Describe the shunt generator, displaying one as you do, and compare it to the series generator. Also, compare both types of shunt generators: self-excited and separately excited. Make sure students can point out the differences and explain why one would be better than the other under certain conditions. Give examples of where shunt generators are used.

Field Excitation Current

Explain what field excitation current is and how it controls the magnetic field, thus controlling the output voltage. Explain what a shunt field rheostat is and how it is used. Also, discuss how the output voltage can be maintained at a constant rate with the use of a voltage regulator. Ask students if they know of a common use of a voltage regulator.

Generator Losses

Explain how voltage drop occurs in the shunt generator, and how the armature is involved in this loss process. Make sure students note that because of the armature's role in voltage drop, an armature with low-resistance is preferred for DC machines.

Discuss the other types of losses, how they occur, and types of corrective measures taken to cut these losses.

29-8 Compound Generators

Explain what a compound generator is, and then explain the two ways the series and shunt fields can be connected. Display a compound generator for students to observe.

29-9 Compounding

Explain what compounding means. Discuss how a generator can become overcompounded, flat-compounded, or undercompounded.

Controlling Compounding

Explain why most DC machines start out at a state of being overcompounded, and how the shunt rheostat or series field diverter is used to reduce the amount of compounding.

Cumulative and Differential Compounding

Explain the two types of compounding and discuss why cumulative compounding is more widely used than differential compounding.

29-10 Countertorque

Explain how countertorque is created by the opposing polarity of the armature and pole pieces. Be sure students note that countertorque is a measure of the useful electrical energy produced by the generator. Explain what dynamic braking or regenerative braking is and how it is used.

29-11 Armature Reaction

Explain what armature reaction is and how it is produced. Have students discuss why armature reaction is a problem.

Correcting Armature Reaction

Discuss the various ways that armature reaction can be corrected. Discuss the advantages and disadvantages of rotating the brushes and of inserting interpoles.

29-12 Setting the Neutral Plane

Using an AC voltmeter and a DC machine, explain and demonstrate how to determine the setting of the brushes and how to set them at the neutral plane. Let students practice doing this.

29-13 Paralleling Generators

Explain why more than one generator may be needed in a given situation, and how an equalizing connection is used. Discuss why the generators are hooked up in parallel, and have students explain why they aren't hooked up in series. Discuss what could happen if an equalizing connection was not made. Demonstrate this setup using two generators and an equalizing connection.

Unit Round Up

Go over the summary in class, demonstrating again any visual points you made during the unit. Clear up any confusion before assigning the review questions and practice problems. Go over these at your next class session, having students show work on the board while they explain their answers. If anything needs to be gone over again, do so before moving on to Unit 30.

UNIT 30
DIRECT CURRENT MOTORS

OUTLINE

KEY TERMS

Brushless DC motors
Compound motor
Constant speed motors
Counter-EMF (CEMF) (back-EMF)
Cumulative-compounded motors
Differential-compounded motors
Field loss relay
Motor
Permanent magnet motors
Printed circuit motor
Pulse-width modulation
Series motor
ServoDisk motor
Servomotors
Shunt motor
Speed regulation
Torque

Anticipatory Set

Go over the unit outline and key terms. Emphasize that this unit will simply go a step further with what was learned in Unit 29. Discuss the various uses of DC motors and ask students for possible examples, correcting them if they suggest a motor that isn't DC. Review the basic principles of repulsion and attraction from Units 1 and 2.

30-1 DC Motor Principles

Compare DC motors with DC generators, pointing out how similar they are. Explain how the DC motor functions, referring to Figure 30-1.

Torque

Define what torque is and describe how it is formed by the magnetic field of the pole pieces and the magnetic field of the loop or armature. Discuss the factors that determine how much torque is produced.

Increasing the Number of Loops

Discuss the advantage of armatures constructed with many turns of wire per loop and many loops.

The Commutator

Compare the use of the commutator in a DC generator and its use in a DC motor. Emphasize that in the DC motor, it forces the direction of current flow to remain constant through sections of the rotating armature.

30-2 Shunt Motors

Describe the construction of the shunt motor and discuss why it is such a good motor to use when constant speed is required.

Counter-EMF

Explain what CEMF is and what three factors it is proportional to.

Speed-Torque Characteristics

Discuss what happens when a DC motor is started without a load. Then discuss what happens when a load is added to the motor. Be sure to talk about the level of torque needed to overcome the various types of losses.

Speed Regulation

Have students write the speed regulation rule into their notes. Be sure that students understand why a lower armature resistance will result in better speed regulation.

30-3 Series Motors

Describe the DC series motor and compare the characteristics of the series motor with those of the DC shunt motor. Be sure students note the relationship of torque to current in the series motor.

Series Motor Speed Characteristics

Emphasize the fact that series motors have no speed limitations and, therefore, need to be coupled with a load at all times. Explain what can happen if a series motor is run without a load. Give examples of where series motors are used.

30-4 Compound Motors

Explain that just like with the compound generator, the compound motor is a combination of the shunt and series motors. Discuss the benefits gained by utilizing the best of each motor, which makes the compound motor the most widely used in industry. Go through the steps for setting up a compound motor, and emphasize how important it is to not set up a differential-compounded motor. Demonstrate how to check for rotation direction on a display motor.

30-5 Terminal Identification for DC Motors

Referring to Figure 30-12, and displaying a DC motor, guide students through the various terminal identifications.

30-6 Determining the Direction of Rotation of a DC Motor

Explain how reversing the connections of the armature leads or field leads changes the rotation direction of the motor. Also, discuss how this is done differently in a compound motor.

30-7 Speed Control

Discuss how reducing the armature current, resulting in less torque, causes the speed to reduce. Then, describe ways of reducing the armature. Also, discuss why some ways of reducing the armature are not the best.

30-8 The Field Loss Relay

Explain how the field loss relay is used to protect the compound motor and the load. Also, discuss the use of two separate shunt fields in compound motors. Have students note that most large DC motors have voltage applied to the shunt field at all times to prevent moisture build-up.

30-9 Horsepower

Review the basic units of power and have students add these to their notes, along with the horsepower formula. Work out the two sample problems on the board or overhead.

30-10 Brushless DC Motors

Describe this motor and compare it to previously studied motors. Explain the use of the converter in supplying power to the brushless motor, and discuss the difference between sine waves and trapezoidal waves. Explain the use of multiple stator poles to achieve a low speed and high torque output.

Inside-Out Motors

Describe and display a rotor wound inside. Make sure students note that motors with a large amount of inertia exhibit superior speed regulation.

Differences between Brush Type and Brushless Motors

Discuss the advantages and disadvantages of each type of motor.

30-11 Converters

Elaborate on the use of converters.

30-12 Permanent Magnet Motors

Explain the make-up of a permanent magnet motor and discuss why it is more efficient than a field wound motor.

Operating Characteristics

Discuss how PM motors function, and compare them to separately excited shunt motors.

DC Servomotors

Describe what servomotors are and how they are used.

DC ServoDisk Motors

Describe this printed circuit motor and compare it to a servomotor, emphasizing the difference in the armature. Explain how the torque and tangential forces are produced.

Characteristics of ServoDisc Motors

Explain why the absence of iron in the armature of the disc motor eliminates the cogging effect that the PM DC motors experience. Discuss pulse-width modulation and how it is used.

30-13 The Right-Hand Motor Rule

Describe how to use the right hand to determine the direction of flow of thrust, field flux, and current through the armature, just as you did previously with the left hand. Compare this to that left-hand rule, and make sure students don't get the two confused.

Unit Round Up

Go over the summary together, checking for understanding, as you have students elaborate on each point in the summary. Have students complete the review questions on their own, and go over these during your next class session.

SECTION 12
Alternating Current Machines

UNIT 31
THREE-PHASE ALTERNATORS

OUTLINE

KEY TERMS

Alternators   Frequency   Revolving field
Brushless exciter   Hydrogen   Rotor
Cooling   Parallel alternators   Sliprings
Excitation current   Phase rotation   Stator
Field discharge resistor   Revolving armature   Synchroscope

Anticipatory Set

As always, make available some devices for some hands-on experience and practice as you progress through the unit. Go over the unit outline and all key terms. Remind students that they have covered three-phase power, and much of this unit will simply add to the knowledge they already have. Do a brief review of frequency from Units 15 and 20.

31-1 Three-Phase Alternators

Give a brief explanation of an alternator. Compare it to the DC generator, and mention the two basic types of alternators.

Revolving Armature Type Alternators

Describe this type of alternator, emphasizing the use of sliprings instead of a commutator. Also, explain why this type is the least used alternator type.

Revolving Field Type Alternators

Describe this alternator and point out the advantage of the stator and rotating magnetic field. Explain how the windings are set up in a wye connection, thus increasing the efficiency of the alternator.

31-2 The Rotor

Explain how the rotor works and display any examples you have, or refer to Figures 31-6 and 31-7.

31-3 The Brushless Exciter

Explain how the brushless alternator is created. Also, be sure students understand that by changing the amount of excitation current to the electromagnets, the amount of voltage induced in the rotor can be varied. Refer to Figures 31-8 through 31-10, or display one of your own during this explanation process.

31-4 Alternator Cooling

Describe how the smaller alternator is cooled, and ask students where they see this type of alternator. Then describe the larger alternator, and explain why using hydrogen is such a good way to keep the large alternator cool.

31-5 Frequency

Explain how frequency is determined and take students through the solution process. Have students add these symbols and this formula into their notes.

31-6 Output Voltage

Discuss the factors that determine output voltage and take students through the formula. Make sure they add these symbols and this formula into their notes. Also, emphasize the rule concerning the length of the conductor.

Controlling Output Voltage

Explain how changing the strength of the magnetic field determines the output voltage.

31-7 Paralleling Alternators

Discuss why two alternators may be needed, and then explain why they must be set up in parallel. Be sure students note how everything must line up, or be in sequence.

Determining Phase Rotation

Explain how to use the lights to determine if both alternators are operating in the same phase rotation. Point out what must be done if the two alternators are not in phase with each other.

Discuss determining the synchronization of the positive peak of phase rotation, and what to do if synchronization hasn't been achieved.

The Synchroscope

If you have a synchroscope, demonstrate how it is used to parallel two alternators. Explain this process while showing the process to students. Also, show students how to use an AC voltmeter to parallel two alternators.

31-8 Sharing the Load

Explain how the load is shared by the two alternators. Have students explain why it was necessary for everything to be synchronized in this sharing the load function.

31-9 Field Discharge Protection

Explain the use of the field discharge resistor in preventing arcing and rotor damage. Discuss exactly how the resistor works. Also, discuss the use of a diode, connected in parallel with the field, as an alternative method of guarding against arcing and damage. Refer to Figures 31-17 through 31-20 as you explain both of these methods.

Unit Round Up

Go through the summary, and allow students to have as much hands-on time with each of the devices as is possible. Assign the review questions as homework, and then have students answer the questions as a group, checking for understanding.

UNIT 32
THREE-PHASE MOTORS

OUTLINE

KEY TERMS

Amortisseur winding
Code letter
Consequent pole
Differential selsyn
Direction of rotation
Dual-voltage motors
Percent slip

Phase rotation meter
Rotating magnetic field
Rotating transformers
Rotor frequency
Selsyn motors
Single-phasing
Squirrel cage

Synchronous
Synchronous condenser
Synchronous speed
Torque
Wound rotor

Anticipatory Set

Go over the unit outline and all key terms. Let students know that everything they have learned thus far will be used in these last two units. They should be familiar with terms like single-phase, three-phase, synchronous, frequency, torque, windings, poles, and motors. Display various types of three-phase motors. Have a phase rotation meter available for student use during that section of the unit.

32-1 Three-Phase Motors

Briefly preview the three types of three-phase motors, distinguishing them from one another by their rotors. Have students note the stator winding as the motor primary, and the rotor as the motor secondary.

32-2 The Rotating Magnetic Field

Discuss the three factors that determine the rotating magnetic field. Refer to Figures 32-1 through 32-8 as you elaborate on the polarity of the voltages and the significance of the arrangement of the stator windings inside the motor. Ask questions about the rotation progress being made inside the motor, and how that is affecting polarity.

Synchronous Speed

Discuss the factors that determine synchronous speed and go over the formula used to calculate synchronous speed.

Determining the Direction of Rotation for Three-Phase Motors

Explain that the direction of rotation of a three-phase motor can be changed by reversing two of its stator leads, and that the direction of rotation can be determined without actually allowing the motor to rotate. Demonstrate how to use a phase rotation meter for this purpose, along with a motor that you can turn the shaft of manually. Be sure students understand how to respond to what registers on the meter.

32-3 Connecting Dual-Voltage Three-Phase Motors

Explain what is meant by a dual-voltage motor, and under what circumstances this type of motor may be required. Then, referring to Figure 32-13, explain the lead numbering process for both wye and delta connections.

High-Voltage Connections

Explain why a high-voltage connection may be necessary, and draw both the wye and delta numbering processes for high voltage on the board or overhead as you explain them. Be sure students realize that the high-voltage connections must be in series.

Low-Voltage Connections

Draw these on the board or overhead as well, explaining where low-voltage connections would be required. Make sure students realize that these connections are in parallel. Also, discuss the adjustments that must be made when a dual-voltage motor has 12 T leads instead of nine.

Voltage and Current Relationships for Dual-Voltage Motors

Discuss the difference that occurs in current flow when a motor is connected to the higher voltage instead of the lower. Review the formulas to solve for phase voltage, line current, total impedance, and phase current.

32-4 Squirrel-Cage Induction Motors

Describe this type of rotor, and display one. Discuss the advantages of this type of rotor.

Principles of Operation

Explain how the squirrel-cage motor works, beginning with the creation of the induced voltage. Discuss how the amount of this induced voltage is determined.

Torque

Make sure students copy the necessary symbols into their notes. Also, be sure students understand why an induction motor can never reach synchronous speed. Have students explain this concept back to you. Also, discuss the significance of the stator and rotor flux being in phase with each other.

Starting Characteristics

Explain what the factors are that will determine the starting current. Display the nameplate of a squirrel-cage motor, and make sure students can find the needed information on it. Take students through the solution process.

Percent Slip

Explain what percent slip is and how to calculate it. Also explain why squirrel-cage motors are considered constant speed motors with minimal percent slip.

Rotor Frequency

Explain what rotor frequency is, how it is calculated, and how it affects rotor bar inductiveness.

Reduced Voltage Starting

Explain why a high starting current is sometimes required with squirrel-cage motors, and why this makes it necessary to reduce the voltage during start-up. Discuss the effect this has on magnetic fields, as well as on the torque.

Code Letters

Explain that the code letters indicate the type of bars used in a given squirrel-cage motor. Discuss what the various letters mean as they pertain to resistance levels. Point out the ripple effect of a particular rating on torque, phase, current, induced voltage, and the rotor magnetic field.

The Double Squirrel-Cage Rotor

Describe and display this type of rotor. Explain the benefits of having the low-resistance bars deep in the core, and high-resistance bars closer to the surface. Have students explain why this type of motor has good running torque and excellent speed regulation.

Power Factor of a Squirrel-Cage Induction Motor

Explain what determines the power factor and how it changes from no load to adding a load.

Single-Phasing

Explain what single-phasing is and how it occurs. Then, discuss what can happen to the motor as a result of single-phasing.

The Nameplate

Display some electric motors, and allow students the opportunity to read the nameplates, and explain what they can do with the information on a nameplate.

Consequent Pole Squirrel-Cage Motors

Explain the difference between these motors and regular squirrel-cage motors. Discuss the advantages of being able to change the synchronous speed, and where these motors would be used. Go through the various connections, ensuring student understanding along the way.

Motor Calculations

Using a wattmeter, demonstrate how to calculate horsepower and motor efficiency. Take students through the solution processes for both.

32-5 Wound Rotor Induction Motors

Describe and display this motor, explaining why its high starting torque and low starting current make it so widely used in industry. Compare the rotors of this motor and the squirrel-cage motor.

Principles of Operation

Explain how this motor functions, beginning with the power applied to the stator winding, and following the process through to the point where all external resistance has been removed from the rotor circuit. Point out that once maximum operating speed is reached, the wound rotor induction motor functions like the squirrel-cage motor.

Starting Characteristics of a Wound Rotor Motor

Compare the starting characteristics of this motor to those of the squirrel-cage motor. Emphasize the difference in starting torque for the two motors, and why it is usually higher in the wound rotor motor.

Speed Control

Explain how speed control takes place and what effect it has on the system.

32-6 Synchronous Motors

Distinguish this motor from other three-phase motors by pointing out its different characteristics.

Rotor Construction

Describe and display the rotor, pointing out how similar it is to the rotor of an alternator.

Starting a Synchronous Motor

Referring to Figure 32-47, explain what the amortisseur winding is, and how this winding is used to start the motor. Also discuss how the windings become electromagnets, allowing the rotor to lock in step with the stator. Make sure that students note, however, that a synchronous motor should never be started with DC current connected to the rotor. Explain what can happen if this occurs.

The Field Discharge Resistor

Explain this resistor's location, and its function in protecting the rotor.

Constant Speed Operation

Explain how this motor starts out as an induction motor, yet doesn't operate as one. Also, explain what pullout torque is, and what happens when pullout torque is reached.

The Power Supply

Discuss the methods that can be used to supply power to a synchronous motor.

Power Factor Correction

Explain how this correction process takes place as the amount of excitation current in the rotor changes.

Interaction of the Direct and Alternating Current Fields

Explain the relationship of the AC current to the DC current in making up for weak DC current, and in producing the torque necessary to operate the load. Also, discuss how the AC current reacts when the DC current is overexcited.

Synchronous Motor Applications

Explain why these motors are widely used in industry, and what types of functions they have.

Advantages of the Synchronous Condenser

Explain the role of the condenser as a power factor correction tool. Discuss why it is better than a bank of capacitors for the same purpose.

32-7 Selsyn Motors

Explain what a selsyn motor is and what it is used for. Display one, if possible. Point out that when these motors are used, at least two are used together. Discuss the reasoning behind this.

Selsyn Motor Operation

Compare the selsyn motor operating process to that of a transformer, as students should already be familiar with transformers. Discuss the rotor position of the transmitter, and how it affects the rotor of the receiver.

The Differential Selsyn

Explain the construction of the differential selsyn, and discuss how it is used.

Unit Round Up

Go through the summary together, having students lead the discussion on each summarized point. Check for any confusion during this time. Have students do the review questions on their own, and then, as a class, discuss them at your next class session. Review any key terms that were not covered in the summary.

UNIT 33
SINGLE-PHASE MOTORS

OUTLINE

KEY TERMS

Centrifugal switch
Compensating winding
Conductive compensation
Consequent pole motor
Holtz motor
Inductive compensation
Multispeed motors

Neutral plane
Repulsion motor
Run winding
Shaded-pole induction motor
Shading coil
Split-phase motors
Start winding

Stepping motors
Synchronous motors
Synchronous speed
Two-phase
Universal motor
Warren motor

Anticipatory Set

Tell students that this final unit will require them to put everything they have learned to actual use in understanding how the various motors work. Almost all of the terms in this unit will be familiar to them. Do a brief review of single-phase power from Unit 26. Display as many of the motors covered in this unit as is possible. Refer to them often as you explain, and demonstrate how they work and where they are used.

33-1 Single-Phase Motors

Stress the fact that single-phase motors have a variety of operating principles, unlike the three-phase motors just studied.

33-2 Split-Phase Motors

Describe the three categories that split-phase motors fall into. Explain how split-phase motors simulate a two-phase power system.

The Two-Phase System

Explain how a two-phase system is created and why it is created.

Stator Windings

Distinguish between the start winding and the run winding. Discuss the purpose of each of these windings and how they function in creating synchronous speed. Refer to Figures 33-2A and 33-2B as you have this discussion.

33-3 Resistance-Start Induction-Run Motors

Discuss torque in this motor, and the factors involved in determining torque. Also, discuss the difference in inductivity between the start winding and the run winding, resulting in phase differentials.

Disconnecting the Start Winding

Explain why the start winding needs to be disconnected. Explain that this is done by centrifugal force in a hermetically sealed unit.

Starting Relays

Explain why starting relays are used on hermetically sealed motors. Describe and display the three types of starting relays and describe on how each works and is used. Refer to Figures 33-8 through 33-12 as you explain the function and operation of each of these relays. Compare the three, and explain which circumstances require which relays.

Relationship of Stator and Rotor Fields

Describe the rotor of the split-phase motor and its function. Explain what takes place after the centrifugal switch opens.

Direction of Rotation

Describe the relationship of the rotation of the rotating magnetic field to the direction of rotation for the motor.

33-4 Capacitor-Start Induction-Run Motors

Compare this motor to the resistance-start induction-run motor. Distinguish the starting characteristics of the two motors. Have students note that the motor should not be started more than approximately eight times per hour, and explain why this is so.

33-5 Dual-Voltage Split-Phase Motors

Describe and display a dual-voltage split-phase motor. Explain how it functions for 120V and how it functions for 240V. Refer to Figure 33-19 to familiarize students with the schematic for this motor's operation.

33-6 Determining the Direction of Rotation for Split-Phase Motors

Explain the connection pattern for both clockwise and counterclockwise rotation.

33-7 Capacitor-Start Capacitor-Run Motors

Explain the differences between these motors and the resistance-start induction-run motors. Point out why a centrifugal switch is not required to disconnect the capacitor from the start winding. Discuss the use of the second capacitor, and describe how it is used in a central air conditioning unit.

Explain what the potential starting relay is and how it is used. Refer to Figure 33-30 as you give this explanation.

33-8 Shaded-Pole Induction Motors

Describe and display this motor.

The Shading Coil

Explain what the shading coil is and how the short circuit effect of the large loop allows a large amount of current to flow. Give examples of motors that use a shading coil. Discuss what happens to the magnetic flux as the AC current reaches its peak and then drops back down to zero.

Speed

Discuss the factors that determine the speed of the shaded-pole induction motor.

General Operating Characteristics

Display a shaded-pole motor, and describe how it functions. Point out that the direction of rotation is determined by the direction of the magnetic field moving across the pole face.

33-9 Multispeed Motors

Describe and display the consequent pole single-phase motor. Also, make sure students realize that the shaded-pole induction motor is also a multispeed motor. Give examples of where these are used.

Multispeed Fan Motors

Referring to Figure 33-37, explain how the run winding has been set up and tapped. Be sure students note that the start winding is set up in parallel with the run winding. Use the air conditioner in a car as an example, as you describe what happens at each setting.

33-10 Repulsion Type Motors

Give a brief explanation of each type of repulsion motor.

33-11 Construction of Repulsion Motors

Display and describe how a repulsion motor is built. Emphasize the difference in the concept of like magnetic poles repelling each other, versus creating a magnetic field.

Operation

Explain how the magnetic field created in the poles is repelled by the magnetic field created in the armature. Have students note that repulsion motors will have the same number of brushes as there are stator poles.

Brush Position

Display this, if possible, as you explain it. Describe what happens, or doesn't happen, when the brushes are at $90°$ to the poles, and when the brushes are lined up with the poles. Then, explain the effect of moving the brushes either clockwise or counterclockwise $15°$.

33-12 Repulsion-Start Induction-Run Motors

Display and describe the two types of repulsion-start induction-run motors. Describe how the brush-riding type functions, and compare it to a squirrel-cage motor. Explain how the brush-lifting type of motor operates, focusing on the role of the weights in the operating process.

33-13 Repulsion-Induction Motors

Describe and display this motor, comparing its workings to those of the repulsion motor and the DC compound motor. Guide students through the schematic in Figure 33-47.

33-14 Single-Phase Synchronous Motors

Explain how these motors work and where they are used. Be sure students realize that constant speed is required, and no DC excitation current is needed.

Warren Motors

Display, if available, and describe the Warren motor. If one is not available, refer to Figure 33-48. Compare this motor to the Holtz motor, noting the difference in the rotor.

Holtz Motor

Display, or refer to Figure 33-49, and describe the construction and operation of this motor.

33-15 Stepping Motors

Display and describe how different these motors are from the other motors studied, especially the Warren and Holtz motors.

Theory of Operation

Referring to Figures 33-50 through 33-53, describe how the stepping motor functions. Discuss the variability of the number of stator poles these motors can have. Have students explain why most stepping motors have eight stator poles.

Windings

Describe the three-lead motor and compare this type of stepping motor to the bifilar stepping motor. Explain the advantage of the two windings over the three-lead type.

Four-Step Switching (Full Stepping)

Explain this process and discuss where students will encounter this in the field.

Eight-Step Switching (Half-Stepping)

Explain this process and compare it to the four-step switching process. Discuss the advantages and disadvantages of each process, in relation to job needs.

AC Operation

Explain why stepping motors operating on AC current become permanent magnet induction motors. Display, or refer to Figure 33-60, as you describe the construction and functioning of the AC operation of this motor.

Stepping Motor Characteristics

Explain the versatility of these motors, and have students discuss why their characteristics are advantageous in a work setting.

33-16 Universal Motors

Compare the universal motor to the DC series motor, but emphasize the major difference of the compensating winding. Describe how the compensating winding winds around the stator, and how it counteracts the inductive reactance in the armature. Have students explain the advantages of having a compensating winding.

Connecting the Compensating Winding for AC Current

Discuss both the conductive compensation connection and the inductive compensation connection. Refer to Figures 33-65 and 33-66 to show how this looks.

The Neutral Plane

Explain how to set the brushes at the neutral plane, using the voltmeter connected to either the series winding or the compensating winding.

Speed Regulation

Explain why the speed regulation of this motor is so poor, and give examples of its use.

Changing the Direction of Rotation

Explain how to change the armature leads with respect to the field leads.

Unit Round Up

Go over the summary together and any key terms not covered in the summary. Have students do review questions on their own, lead the discussions covering these questions at your next class session. Listen carefully, to check for any area that appears to be unclear, and go back over those areas if there are any.

Answers to Review Questions

Unit 1
ATOMIC STRUCTURE

1. Neutron (No charge)
   Proton (+)
   Electron (-)
2. 3
3. 1840
4. Opposite charges attract and like charges repel.
5. Centrifugal force
6. 1 or 2
7. 7 or 8
8. The flow of electrons
9. A theoretical particle that bonds quarks together as well as protons and neutrons.
10. Quarks

Unit 2
ELECTRICAL QUANTITIES AND OHM'S LAW

1. A quantity of electrons
2. One coulomb per second
3. Electromotive force, electrical pressure, potential difference
4. A measurement of the resistance to the flow of electricity
5. A measurement of electrical power
6. 7.5 amps. (120 / 16 = 7.5)
7. 900 watts (120 x 7.5 = 900)
8. 12 ohms (240 / 20 = 12)
9. 120 volts (15 x 8 = 120)
10. 30 amps (240 / 8 = 30)
11. 7200 watts (240 x 30 = 7200)
12. 20.83 amps (5000 / 240 = 20.83)
13. 41.67 amps (5000 / 120 = 41.67)
14. Since the power consumption is the same, it will cost the same to operate the heating element regardless of which voltage is used.

Unit 3
STATIC ELECTRICITY

1. Because the electrons are sitting still and not moving
2. Positively charged
3. Because insulators hold the electrons in position and do not let them move to a different position
4. A device used to test the electrostatic charge of a material

5. Positive charge
6. Yes
7. From the cloud to the ground
8. Lightning rods and lightning arrestors
9. Selenium
10. Its conductivity changes with a change in light intensity.

Unit 4
MAGNETISM

1. South polarity
2. Lodestones
3. Repel each other
4. Using the left hand rule
5. Flux density—A measurement of the strength of a magnetic field.
   Permeability—The measure of a material's willingness to become magnetized.
   Reluctance—Resistance to magnetism.
   Saturation—The maximum line of magnetic force a material can hold.
   Coercive force—A material's ability to retain magnetism.
   Residual magnetism—The amount of magnetic force remaining in a piece of material after the magnetizing force has been removed.
6. 27,800 dynes.

Unit 5
RESISTORS

1. Composition carbon, metal film, metal glaze, wire wound
2. They do not change their value with age.
3. Higher power rating to withstand more heat
4. Vertically to permit air to rise through the center
5. Yes (.01 x .01 x 2000 = .2 watts)
6. No (24 x 24 / 350 = 1.645 watts)
7. 360,000 ohms at ±5%
8. 10,500 - 9,500
   (10,000 x 0.05 = 500 ; 10,000 + 500 = 10,500 ; 10,000 - 500 = 9,500)
9. Yes
10. A variable resistor connected to operate as a voltage divider.

Unit 6
SERIES CIRCUITS

1. 707 ohms (200 + 86 + 91 + 180 + 150 = 707 Ω)
2. 119 ohms (360 - 56 - 110 - 75 = 119 Ω)
3. 20 Volts (120 - 35 - 28 - 22 - 15 = 20 volts)
4. 0.2 amp (44V / 220 Ω = 0.2 A)
5. 240 Volts (60 + 60 + 60 + 60 = 240 volts)
6. A circuit which has only one path for current flow
7. 1. The current is the same at any point in the circuit.
   2. The sum of the voltage drops is equal to the applied voltage.
   3. The total resistance is equal to the sum of the individual resistors.
8. 462 ohms (160 + 100 + 82 + 120 = 462 Ω)
9. 0.0519 amp (24 / 462 = 0.0519 A)
10. 160 Ω = 8.304 Volts (160 x 0.0519 = 8.304 V)
    100 Ω = 5.19 Volts (100 x 0.0519 = 5.19 V)
    82 Ω = 4.256 Volts (82 x 0.0519 = 4.256 V)
    120 Ω = 6.228 Volts (120 x 0.0519 = 6.228 V)

Unit 7
PARALLEL CIRCUITS

1. It has more than one path for current flow.
2. Because each device receives the same voltage
3. 1. The voltage is the same across any point in a parallel circuit.
   2. The total current is equal to the sum of the currents flowing through each separate path.
   3. The reciprocal of the total resistance is equal to the sum of the reciprocals of each resistor.
4. 3.75 amps (0.8 + 1.2 + 0.25 + 1.5 = 3.75 A)
5. 25 Ω (R_T = R/N ; R_T = 100/4 ; R_T = 25 Ω)
6. 0.85 Amp (2.8 - 0.9 - 1.05 = 0.85 A)

7. 90.741 Ω ($R_T = \dfrac{1}{\dfrac{1}{R_1} + \dfrac{1}{R_2} + \dfrac{1}{R_3} + \dfrac{1}{R_N}}$)

8. 273.76 Ω ($R_4 = \dfrac{1}{\dfrac{1}{R_T} - \left(\dfrac{1}{R_2} + \dfrac{1}{R_3} + \dfrac{1}{R_N}\right)}$)

9. The first step in solving this problem is to find the total resistance.

$$R_T = \dfrac{1}{\dfrac{1}{R_1} + \dfrac{1}{R_2} + \dfrac{1}{R_3} + \dfrac{1}{R_N}} = 628.571\Omega$$

The second step is to determine the amount of voltage applied to the circuit.

$E = I \times R$; $E = 0.25 \times 628.571 = 157.143$ volts
Now that the voltage drop across each resistor is known, the current flow through each resistor can be determined.
1200 Ω = 0.131 A (157.143/1200 = 0.131)
2200 Ω = 0.0714 A (157.143/2200 = 0.0714)
3300 Ω = 0.0476 A (157.143/3300 = 0.0476)

Unit 8
COMBINATION CIRCUITS

1. R_T = 440.791 Ω To find the total resistance, first combine resistors R_1 & R_2, and R_3 & R_4.
[$R_{C\,(1\&2)}$ = 470 + 360 ; $R_{C\,(1\&2)}$ = 830 Ω]
[$R_{C(3\&4)}$ = 510 + 430 ; $R_{C(3\&4)}$ = 940 Ω] Since these combination values are connected in parallel, the total resistance can be computed using a parallel resistance formula:

$$R_T = \dfrac{1}{\dfrac{1}{R_1} + \dfrac{1}{R_2} + \dfrac{1}{R_3} + \dfrac{1}{R_N}} R_T = 440.791\Omega$$

E_T = 264.475 Volts After the total resistance has been found, the total voltage can be computed using Ohm's law. ($E_T = I_T \times R_T$; $E_T = 0.6 \times$ 440.791; E_T = 264.475 volts)
I_1 and I_2 = 0.319 Amp The current flow through the branch containing resistors R_1 and R_2 can be computed using the combined resistance and the applied voltage. ($I_{C(1\&2)}$ = 264.475/830; $I_{C(1\&2)}$ = 0.319 A). Since resistors R_1 and R_2 are connected in series, each has the same current flow.
E_1 = 149.93 Volts The voltage drop across resistors R_1 and R_2 can be computed using Ohm's law. (E_1 = 0.319 x 470; E_1 = 149.93 volts)
E_2 = 114.84 Volts (E_2 = 0.319 x 360; E_2 = 114.84 volts)
I_3 and I_4 = 0.281 Amp The current flow through the branch containing resistors R_3 and R_4 can be computed using the combined resistance and the applied voltage. $I_{C(1\&2)}$ = 264.475/940; $I_{C(3\&4)}$ = 0.281 A. Since resistors R_3 and R_4 are connected in series, each has the same current flow.
E_3 = 143.31 Volts (E_3 = 0.281 x 510; E_3 = 143.31 volts).
E_4 = 120.83 Volts (E_4 = 0.281 x 430; E_4 = 120.83 volts).

2. R_T = 5627.105 Ω To determine total resistance, the parallel blocks containing resistors R_2 & R_3 and R_5 & R_6 must be combined into one single resistor.

$1/R_{C(2\&3)} = 1/2200 + 1/1800$; $R_{C(2\&3)} = 990$ Ω

$1/R_{C(5\&6)} = 1/3300 + 1/4300$; $R_{C(5\&6)} = 1867.105$ Ω

The combined resistance values of $R_{C(2\&3)}$ and $R_{C(5\&6)}$ are connected in series with resistors R_1, R_4, and R_7.

$R_T = R_1 + R_{C(2\&3)} + R_4 + R_{C(5\&6)} + R_7$; $R_T = 1000 + 990 + 910 + 1867.105 + 860$

R_T = 5627.105 Ω

I_T, I_1, I_4, and I_7 = 0.0112 Amps The total resistance and total voltage can be used to compute total current using Ohm's Law. $I_T = E_T / R_T$; I_T = 63 / 5627.105.

I_T = 0.0112 amp

Since resistors R_1, R_4, and R_7 are single resistors connected in series, they will have the same current flow.

E_1 = 11.2 Volts ($E_1 = I_1 \times R_1$; E_1 = 0.0112 x 1000)

E_2 and E_3 = 11.088 Volts Resistors R_2 and R_3 are connected in parallel. The voltage dropped across each can be computed using the combined resistance and the current flow through the block.

I_2 = 0.00504 Amp (11.088 / 2200 = 0.00504 A)

I_3 = 0.00616 Amp (11.088 / 1800 = 0.00616 A)

E_4 = 10.192 Volts (0.0112 x 910 = 10.192)

E_5 and E_6 = 20.912 Volts

E_5 & $E_6 = I_T \times R_{C(5\&6)}$; E_5 & E_6 = 0.0112 x 1867.105; E_5 & E_6 = 20.912 volts

I_5 = 0.00634 Amp (20.912 / 3300)

I_6 = 0.00486 Amp (20.912 / 4300)

E_7 = 9.632 Volts (0.0112 x 860)

3. The algebraic sum of the voltage sources and voltage drops in a closed circuit must equal zero.

4. The algebraic sum of the currents entering and leaving a point must equal zero.

5. To reduce the circuit to a single power source and series or parallel resistance

6. Positive

7. The assumed direction of current flow was incorrect.

8. I_1 = 0.0521 Amp

$-E_1 - E_3 = -24$ $E_1 + E_3 = 24$

$-E_2 - E_3 = -10$ $E_2 + E_3 = 10$

$E_1 = I_1 \times R_1$ $E_1 = I_1 \times 330$ $E_1 = 330 I_1$

$E_2 = I_2 \times R_2$ $E_2 = I_2 \times 200$ $E_2 = 200 I_2$

$E_3 = (I_1 + I_2) \times R_3$ $E_3 = (I_1 + I_2) \times 100$ $E_3 = 100(I_1 + I_2)$

$330 I_1 + 100 I_1 + 100 I_2 = 24$

$200 I_2 + 100 I_1 + 100 I_2 = 10$

$430 I_1 + 100 I_2 = 24$

$100 I_1 + 300 I_2 = 10$

$-3(430 I_1 + 100 I_2 = 24)$ $-1290 I_1 - 300 I_2 = -72$

$-1290 I_1 - 300 I_2 = -72$

$100 I_1 + 300 I_2 = 10$

$-1190 I_1 = -62$

$-62 / -1190 = 0.0521$ A

9. I_2 = 0.0160 amp

$100 I_1 + 300 I_2 = 10$

$(100 \times 0.0521) = 300 I_2 = 10$

$300 I_2 = 10 - 5.21$ $300 I_2 = 4.79$ $I_2 = 0.0160$ A

10. E_1 = 17.193 volts (0.0512 x 330)

11. E_2 = 3.2 volts (0.0160 x 200)

12. E_3 = 6.81 volts

$I_3 = I_1 + I_2$ $I_3 = 0.0521 + 0.0160$ $I_3 = 0.0681$ A

E_3 = 0.0681 x 100

13. E_{thev} = 40 volts (4 + 20 = 24) (48 / 24 = 2 A) (2 x 20 = 40 volts)

14. R_{thev} = 3.333 ohms ($1/R_{thev} = 1/R_1 + 1/R_2$)

15. I_{norton} = 8 amps (20 / 2.5)

16. R_{norton} = 2.162 ohms ($1/R_{Norton} = 1/R_1 + 1/R_2$)

17. I_{load} = 3.529 amps (3.333 + 8 = 11.333; 40 / 11.333 = 3.529 A)

Unit 9
MEASURING INSTRUMENTS

1. The amount of current flow through the moving coil.

2. DC

3. 5 Meg. Ω (20,000 x 250)

4. To give the circuit current another path to flow, bypassing the meter movement

5. 1. Use a make-before-break switch
   2. Ayrton shunt

6. In series

7. In parallel

8. 0.001 Ω (0.05 / 50)
9. Ohmmeter
10. Voltage
11. Amplitude of voltage
12. Time
13. 20K Hz. (1/0.000050)
14. The D'Arsonval movement contains a permanent magnet. The wattmeter contains an electromagnet connected in series with the load.
15. The amount of current flow in the circuit and the amount of voltage.

Unit 10
USING WIRE TABLES AND DETERMINING CONDUCTOR SIZES

1. 75°C or 167°F (310-13)
2. UF and USE (310-13)
3. 44 amps 50 x 0.88 = 44 (310-17)
4. 96 amps 120 x 0.80 = 96 (310-16) (Note 8a)
5. A. Be the same length

   B. Have the same conductor material

   C. Be the same size in circular mil area

   D. Have the same insulation type

   E. Be terminated in the same manner
6. #10 AWG (310-3)
7. No (310-13)
8. External ridge (310-12a)
9. White, neutral gray, or green (310-12c)
10. 17.28 amps (30 x 0.96 x 0.60 = 17.28) (310-16) (Note 8a)
11. Max. volt drop (480 x 0.03 = 14.4 volts)

    Max. wire resistance (14.4 / 86 = 0.167Ω)

    Length of conductor (2800 x 2 = 5600 ft.)

    Conductor size (17 x 5600 / 0.167 = 570,059.88 CM)

    ANSWER: 600 kcmil cable
12. 3 (570,059.88 / 211,600 = 2.7)
13. Max. voltage drop (480 x 0.06 = 28.8 volts)

    Max. wire resistance (28.8 / 235 = 0.122 Ω)

    Length of conductor (1800 x 2 x 0.866 = 3117.6 ft.)

    Wire size (10.4 x 3117.6 / 0.122 = 265,762.6 CM)

    ANSWER: 300 kcmil cable

Unit 11
CONDUCTION IN LIQUIDS AND GASES

1. Ions
2. An atom which has gained one or more electrons
3. An atom which has lost one or more electrons
4. Acids, alkalies, and metallic salts
5. A liquid solution which permits the flow of electric current through it
6. Two extra electrons
7. The cathode
8. The anode
9. The length of time the process is permitted to continue
10. The process of separating elements electrically
11. The amount of voltage required to start conduction in a gas
12. Electron impact
13. 1. The applied voltage

    2. The distance traveled before striking a gas molecule
14. The pressure inside the tube
15. The type of gas in the tube

Unit 12
BATTERIES AND OTHER SOURCES OF ELECTRICITY

1. A device which converts chemical energy into electrical energy
2. The materials it is made of
3. The area of the cell plates
4. A group of cells connected together
5. A cell which can not be recharged
6. A cell which can be recharged.
7. Mercuric oxide
8. Carbon-zinc cell
9. Ammonium chloride, manganese dioxide, and granulated carbon.
10. Longer life
11. Manganese dioxide
12. The amount of current the battery can produce for a period of 20 hours at 80°F
13. Hydrometer
14. 540 amps (180 x 3) 4.8 volts (6 x 0.80) 3 minutes
15. 36 volts 100 AH
16. 0.5 volts
17. The surface area
18. Three banks of 12 cells should be connected in series. Each bank should then be connected in parallel.

19. The type material it is made of and the difference in temperature between the two junctions
20. Types R, S, and B
21. Iron and constantan
22. Pressure

Unit 13
MAGNETIC INDUCTION

1. The direction of current flow
2. The amount of current flow
3. 1. The number of turn of wire
   2. The strength of the magnetic field
   3. The speed of the cutting action
4. 100,000,000
5. It has the effect of adding turns in series causing the induced voltage in each to add.
6. 5
7. 63.2%
8. 0.0417 second
9. 750.6 volts (250.6 x 3)
10. Diode

Unit 14
BASIC TRIGONOMETRY

1. Cosine
2. 33.56 feet
3. 2.55 meters (12 x TAN 12°)
4. 8.48 inches (6 / COS 45°)
5. 28.48° (SIN$\angle$ = 31 / 65)
6. 80.03 feet
7. 57.99° (TAN$\angle$ = 2 / 1.25)
8. 16.007 feet (14 / SIN 61°)
9. 7.76 feet (14 / TAN 61°)
10. 54° (90° - 36°)

Unit 15
ALTERNATING CURRENT

1. Sine wave
2. 360°
3. 270°
4. The number of complete cycles that occur in one second
5. 141.6 volts (230 x SIN 38°)
6. 168.18 volts (63 / SIN 22°)
7. 12.69° (SIN$\angle$ = 123 / 560)
8. 306.13 volts (433 x 0.707)
9. 28.03 volts (88 / 2 x 0.637)
10. 75.55 volts (68 / 0.9)

Unit 16
INDUCTANCE IN ALTERNATING CURRENT CIRCUITS

1. 0° Current and voltage are in phase.
2. 90°
3. Inductance of the inductor and the frequency
4. 2.4 Henrys (0.6 x 4)
5. 0.0214 Henry (1/L_T = 1/0.05 + 1/0.06 + 1/0.1)
6. 79.17 Ω (L_T = 0.06 + 0.05 + 0.1; L_T = .21 H; X_L = 377 x 0.21)
7. 0.737 H [X_L = 400 Ω (240 / 0.6) L = 400 / 2 x π x 1000]
8. 0.354 amp (X_L = 2 x π x 60 x 3.6) (I = 480 / 1357.2)
9. 0.424 amp (X_L = 2 x π x 50 x 3.6) (I = 480 / 1130.973)
10. 1,666.3 Ω (L = 250 / 2 x π x 60) (X_L = 2 x π x 400 x 0.663)

Unit 17
RESISTIVE-INDUCTIVE SERIES CIRCUITS

1. The voltage and current are in phase with each other.
2. The current lags the voltage by 90°.
3. The amount of apparent power as compared to the true power.
4. 40.36 Ω (XL = 377 x 0.093; XL = 35.061Ω;
$$Z = \sqrt{20^2 + 35.061^2})$$
5. 30.68° (COS $\angle\varnothing$ = 0.86)
6. 143.18 Vars ($VARs_L = \sqrt{VA^2 - P^2}$;
$$VARs_L = \sqrt{230^2 - 180^2})$$
7. 59.94 Volts ($E_T = \sqrt{E_R^2 + E_L^2}$;
$$E_T = \sqrt{53^2 + 28^2})$$
8. 16.66°
$$(P = \sqrt{4329^2 - 1234^2} \quad PF = \frac{4149.396}{4329};$$
$$COS\angle\varnothing = 0.958)$$

9.10.09 Ω (Z =120 / 12; Z= 10 Ω;

$$X_L = \sqrt{Z^2 - R^2}\,;\ X_L = \sqrt{10^2 - 6.5^2}\,)$$

10. 78 Volts (12 x 6.5)

Unit 18
RESISTIVE-INDUCTIVE PARALLEL CIRCUITS

1. 90°
2. 2.88 Amps (X_L = 377 x 0.2;

$$Z = \cfrac{1}{\sqrt{\left(\cfrac{1}{50}\right)^2 + \left(\cfrac{1}{75.4}\right)^2}}\,;\ I_T = 120\ /\ 41.67)$$

3. 41.67 Ω (120 / 2.88)
4. 83.3% (41.67 / 50 = 0.833 x 100)
5. 33.59° (COS $\angle\varnothing$ = 0.833)
6. 10.308 Amperes ($I_{TOTAL} = \sqrt{6.5^2 + 8^2}$)
7. 15.364 Ω ($Z = \cfrac{1}{\sqrt{\left(\cfrac{1}{24}\right)^2 + \left(\cfrac{1}{20}\right)^2}}$)
8. 214.5 Watts (325 x 0.66)
9. 337.38 Vars ($VARs_L = \sqrt{465^2 - 320^2}$)
10. 46.528° (PF = 320 / 465; COS $\angle\varnothing$ = 0.688)

Unit 19
CAPACITORS

1. The insulation between the plates
2. a. The area of the plates
   b. The distance between the plates
   c. The type of dielectric
3. 5 (15 / 3 = 5)
4. Electrostatic charge
5. 170 μF (20 + 50 + 40 + 60)
6. 8.955 μF (1/C_T = 1/20 + 1/50 + 1/40 + 1/60)
7. 9.9 seconds (T = 90,000 x 0.000022; T = 1.98 sec, ; 1.98 x 5 = 9.9)
8. 222.222M Ω (T = 0.5 / 5 ; T = 0.1 sec; R = 0.1 / $450^{x10(-12)}$)
9. Yes
10. By connecting two wet type electrolytic capacitors back to back
11. Dry type electrolytic

12. 0.75 second (T = 300,000 x 0.0000005; T = 0.15 x 5)
13. 25,000 pF ±3% (Refer to chart in Figure 19-32)
14. 470 pF ±5% 600 volts (Refer to chart in Figure 19-29)
15. 33 pF ±10% (Refer to chart in Figure 19-29)

Unit 20
CAPACITANCE IN ALTERNATING CURRENT CIRCUITS

1. No
2. Frequency and capacitance of the capacitor
3. 90°
4. Lead
5. 2.714 amps (X_C = 1 / 377 x 0.000030; 240 / 88.419)
6. 10.186 μF (X_C = 1250 / 80 ; C = 1 / 2 x π x 1000 x 15.625)
7. Yes (480 x 1.414 = 678.72 volts)
8. 8 times
9. 17.237 μF (X_C = 277 / 12; C = 1 / 2 x π x 400 x 23.083)
10. Yes (350 x 1.414 = 494.9)

Unit 21
RESISTIVE-CAPACITIVE SERIES CIRCUITS

1. Lead
2. 31.244 Ω (X_C = 1 / 377 x 0.0001105;
$$Z = \sqrt{20^2 + 24.005^2}\,)$$
3. 40.45° (COS $\angle\varnothing$ = 0.76)
4. X_C = 77.666 Ω ($X_c = \sqrt{84^2 - 32^2}$)
5. 3 Ω (C = 1 / 377 x 50; X_C = 1/2 x π x 1000 x 0.00005305)

Unit 22
RESISTIVE-CAPACITIVE PARALLEL CIRCUITS

1. 90°
2. 6.707 amps (X_C = 1 / 377 x 0.0001326;

$$Z = \cfrac{1}{\sqrt{\left(\cfrac{1}{40}\right)^2 + \left(\cfrac{1}{20.004}\right)^2}}$$

I_T = 120 / 17.891)

3. $17.891\ \Omega$ $(Z = \dfrac{1}{\sqrt{\left(\dfrac{1}{40}\right)^2 + \left(\dfrac{1}{20.004}\right)^2}})$

4. 44.727% (Z/R; 17.891 / 40)

5. $63.431°$ (COS $\angle\varnothing$ = 0.44727)

Unit 23
RESISTIVE-INDUCTIVE-CAPACITIVE SERIES CIRCUITS

1. The voltage is in phase with the current.
2. The voltage leads the current by 90°.
3. The voltage lags the current by 90°.
4. $17.067\ \Omega$ (X_L = 2 x π x 400 x 0.0119; X_L = 29.908 Ω

    X_C = 1 / 2 x π x 400 x 0.0000166; X_C = 23.969 Ω;

    $Z = \sqrt{16^2 + (29.908 - 23.969)^2}$)
5. 25.781 amps (440 / 17.067)
6. E_R = 412.496 volts (25.781 x 16)

    E_L = 771.058 volts (25.781 x 29.908)

    E_C = 617.945 volts (25.781 x 23.969)
7. $10,634.456$ watts (412.492 x 25.781)
8. $11,343.64$ VA (440 x 25.781)
9. 93.748% (PF = 10,634.456 / 11,343.64)
10. $20.367°$ (COS $\angle\varnothing$ = 0.93748)

Unit 24
RESISTIVE-INDUCTIVE-CAPACITIVE PARALLEL CIRCUITS

1. X_L = 39.961 Ω (X_L = 2 x π x 400 x 0.0159)

    X_C = 29.916 Ω (X_C = 1/2 x π x 400 x 0.0000133)

    I_R = 10 amps (240 / 24)

    I_L = 6.006 amps (240 / 39.961)

    I_C = 8.022 amps (240 / 29.916)

    P = 2,400 watts (240 x 10)

    $VARS_L$ = 1,441.44 (240 x 6.006)

    $VARS_C$ = 1,925.28 (240 x 8.022)

    I_T = 10.201 amps

    $(I_{TOTAL} = \sqrt{10^2 + (6.006 - 8.022)^2})$

    VA = 2,448.24 (240 x 10.201)

    PF = 98.03 % (PF = 2400 / 2448.24)

    $\angle\varnothing$ = 11.392 ° (COS $\angle\varnothing$ = 0.9803)

2. $10.243\ \Omega$ $(Z = \dfrac{1}{\sqrt{\left(\dfrac{1}{16}\right)^2 + \left(\dfrac{1}{8} - \dfrac{1}{20}\right)^2}})$

3. 40.497 amps $(I_{TOTAL} = \sqrt{38^2 + (22 - 8)^2})$

4. 0.615 amps (I_L & I_C = 277 / 30; I_{RES} = 9.233 / 15)

5. 318.22 nF or 0.31822 µF (X_L = 2 x π x 10,000 x 0.000796; X_L = 50.014 Ω

    C = 1/2 x π x 10,000 x 50.014)

6. PF = 63% (VA = 560 x 53; VA = 29,680; PF = 18,700 / 29,860)

    C = 194.952 µF

    $(VARs_L = \sqrt{29,000^2 - 18,700^2}$; $VARS_L$ = 23,048.045

    23,048.045 capacitive VARs are needed to overcome the inductive VARs.

    $X_c = \dfrac{560^2}{23,048.045}$; X_C = 13.606 Ω; C = 1 / 377 x 13.606)

Unit 25
FILTERS

1. 338.627 Hz. (1/2 x π x 10E-6 x 47)
2. 33.863 Hz. (1/2 x π x 10E-6 x 470)
3. $E_{CAPACITOR}$ = 8.976 volts $E_{RESISTOR}$ = 13.254 volts

    $(X_c = \dfrac{1}{2\pi FC})$ (X_C = 1/2 x π x 500 x 10E-6) (X_C = 31.831 Ω)

    $Z = \sqrt{R^2 + X_c^2}$ $Z = \sqrt{47^2 + 31.831^2}$ Z = 56.764 Ω

    I = E/Z   I = 0.282 A

    $E_{RESISTOR}$ = 47 x 0.282 $E_{RESISTOR}$ = 13.254 volts

    $E_{CAPACITOR}$ = 31.831 x 0.282 $E_{CAPACITOR}$ = 8.976 volts

4. 1027.341 Hz.

    $(F_R = \dfrac{1}{2 \times \pi \times \sqrt{0.0012 \times 0.000020}})$

5. 1027.341 Hz. The resonant frequency is the same for series or parallel resonance.

6. 265.258 Hz. $F = \dfrac{X_L}{2 \times \pi \times L}$

$F = \dfrac{10}{2 \times \pi \times 0.006}$

7. C = 8.443 µf

$X_L = 2\pi FL$ $X_L = 2 \times \pi \times 1000 \times 0.003$ $X_L = 18.850 \ \Omega$

$C = \dfrac{1}{2\pi F X_c}$ $C = \dfrac{1}{2 \times \pi \times 1000 \times 18.850}$

C = 8.443 µf

8. L = 5.874 mh

$X_c = \dfrac{1}{2\pi FC}$ $X_c = \dfrac{1}{2 \times \pi \times 1400 \times 0.0000022}$

$X_C = 51.674 \ \Omega$

$L = \dfrac{X_L}{2\pi F}$ $L = \dfrac{51.674}{2 \times \pi \times 1400}$ L = 0.005874 h

Unit 26
THREE-PHASE CIRCUITS

1. 120°
2. Wye and delta
3. 277.136 volts (480 / 1.732)
4. 25 amps ($I_{Phase} = I_{Line}$)
5. 560 volts ($E_{Phase} = E_{Line}$)
6. 51.96 amps (30 x 1.732)
7. VA = 12,960 (VA = 3 x 240 x 18)
8. E_L = 415.68 volts I_L = 18 amperes (E_L = 240 x 1.732) ($I_L = I_P$)
9. 103.919 Ω (I_{Phase} = 40 / 1.732; Z = 2400 / 23.095)
10. VA = 166,272 (VA = $\sqrt{3}$ x 2400 x 40)

Unit 27
SINGLE-PHASE TRANSFORMERS

1. A machine that can change values of voltage, current, and impedance without changing frequency.
2. 90% to 99%
3. A transformer that has its primary and secondary windings electrically and mechanically separated from each other.
4. Turns ratio
5. A transformer that has only one winding used for both the primary and secondary.
6. It has no line isolation.

7. A step-up transformer has a higher secondary voltage than primary voltage.
   A step-down transformer has a lower secondary voltage than primary voltage.
8. 5:1 (240 / 48)
9. 6.25 amps (750 / 120)
10. 3 amps (18 / 6)
11. The polarity of the windings
12. 160 Volts (240 - 80)
13. 320 Volts (240 + 80)
14. 4

Unit 28
THREE-PHASE TRANSFORMERS

1. Two
2. 173.2 KVA (200 x 0.866)
3. By an orange wire or by tagging
4. 489.76 (277 x 1.732)
5. 12 amps (80 - 68)
6. 208 volts (120 x 1.732)
7. 277 volts (480 / 1.732)
8. 208 volts (120 x 1.732)
9. No
10. Yes
11. 120 Hz.
12. 3rd, 9th, and 15th. A triplen must be an odd multiple of the third harmonic. The 6th and 12th are even multiples.
13. A negative rotating harmonic is more harmful. The negative rotating harmonic causes a reduction in motor torque, which can cause it to slow down and overheat.
14. Harmonic analyzer
15. Yes, it should be derated to 14.6 kVA (94 x 1.414 = 132.916 amps; 132.916/204 = 0.651 - 22.5 kVA x 0.651 = 14.6 kVA)

Unit 29
DIRECT CURRENT GENERATORS

1. A machine that changes mechanical energy into electrical energy.
2. Alternating voltage
3. Lap, wave, and frogleg
4. Wave winding
5. Small pole pieces located between the main pole pieces. They are used to help correct armature reaction.
6. In series
7. Shunt field winding
8. In series

9. In parallel
10. The twisting or bending of the main magnetic field
11. Armature current
12. Currents that are induced in the iron core material
13. The output voltage is higher at full load than it is at no load.
14. To control the output voltage by adjusting the amount of field excitation current.
15. Series field shunt rheostat
16. Cumulative compound—The series and shunt fields aid each other in the production of magnetism.

    Differential compound—The series and shunt fields oppose each other in the production of magnetism.
17. 1. Number of turns of wire in the armature or length of the conductor

    2. Strength of the magnetic field of the pole pieces

    3. Speed of the armature
18. The resistance of the armature
19. Armature current if the field excitation current remains constant
20. The useful electrical energy produced by the generator

Unit 30
DIRECT CURRENT MOTORS

1. Attraction and repulsion of magnetism
2. A machine that converts electrical energy into mechanical energy
3. It performs the function of a rotary switch to keep the current flowing in the proper direction through the armature.
4. 1. The strength of the magnetic field of the pole pieces

   2. The strength of the magnetic field of the armature
5. The shunt motor
6. A voltage induced in the armature windings that opposes the applied voltage
7. 1. The strength of the magnetic field of the pole pieces

   2. The number of turns of wire in the armature

   3. The speed of the armature
8. The wire resistance of the armature
9. Armature resistance
10. Series

11. Differential compound
12. Change the armature leads
13. By connecting full voltage to both the armature and shunt field
14. By connecting full voltage to the armature and reducing the current flow through the shunt field
15. The shunt field relay
16. The resistance of the shunt field produces heat, which prevents the formation of moisture inside the motor.
17. James Watt
18. 746
19. 18.782 HP (1.59 x 750 x 1575 / 100,000)
20. 70% (18.78 x 746 = 14,009.88 watts)

    (250 x 80 = 20,000 watts)

    (14,009.88 / 20,000 = 0.70)

Unit 31
THREE-PHASE ALTERNATORS

1. A. The phase rotation must be the same

   B. The output voltages should be the same

   C. The output voltages must be in phase
2. By reversing the connection of two stator leads
3. A. Indicates phase rotation of the two machines

   B. Indicates phase sequence of the two machines
4. An instrument used for paralleling two alternators, which can determine phase rotation and difference in frequency between the two alternators
5. The input power of alternator B must be increased.
6. Number of stator poles and speed of rotation
7. 1200 RPM
8. A. Length of the conductor of the armature or stator

   B. Strength of the magnetic field of the rotor

   C. Speed of the rotor
9. To supply excitation current to the rotor
10. DC
11. A three-phase bridge rectifier
12. A field discharge resistor or a diode

Unit 32
THREE-PHASE MOTORS

1. A. Squirrel-cage induction
   B. Wound rotor induction
   C. Synchronous
2. Rotating magnetic field
3. The speed of the rotating magnetic field
4. A. Number of stator poles per phase
   B. Frequency of the applied voltage
5. A. The three voltages are 120° out of phase with each other.
   B. The voltages reverse polarity at regular intervals.
   C. The arrangement of the stator windings
6. A. The strength of the magnetic field of the stator
   B. The strength of the magnetic field of the rotor
   C. The phase angle difference between rotor and stator flux
7. No
8. A squirrel-cage winding used to start a synchronous motor
9. The rotor will not turn and it may damage rotor windings and power supply.
10. A. Runs at the speed of the rotating magnetic field
    B. Operates at a constant speed from full load to no load
    C. Can be made to have a leading power factor
11. To prevent excessive voltage from being induced in the rotor winding during starting
12. At synchronous speed there is no voltage induced in the rotor to produce current flow. If there is no current flow there can be no magnetic field to produce torque.
13. 4.919 Hp. (6.28 x 1175 x 22 / 33000)
14. 64.95% (4.919 x 746 / 5650 = 0.6495)
15. The wound rotor motor contains a rotor with windings and sliprings that provide connection to an external circuit.
16. Higher starting torque, lower starting current, and speed control
17. A. The strength of the magnetic field of the stator
    B. The number of turns of wire in the rotor
    C. The speed of the cutting action
18. There is no path for current flow and a magnetic field can not be developed in the rotor.
19. The resistance connected to the rotor circuit causes the rotor current and induced voltage to be more in phase with each other. This causes

the rotor and stator flux to be more in phase, resulting in a higher starting torque.
20. When it is operated without load and used for the purpose of power factor correction only
21. The power factor will be 100% and unity and the AC current applied to the motor will be at its lowest value.
22. By over-excitation of the field.
23. DC
24. By changing the number of stator poles
25. 8 (synchronous speed of 900 RPM)

Unit 33
SINGLE-PHASE MOTORS

1. A. Resistance-start induction-run
   B. Capacitor-start induction-run
   C. Capacitor-start capacitor-run
2. 90°
3. In parallel
4. 90°
5. It develops higher starting torque.
6. 35° to 40°
7. A centrifugal switch
8. A rotating magnetic field is produced by the stator and rotor.
9. By reversing the connection of the start winding leads in relation to the run winding
10. In series
11. From the rear
12. Capacitor-start capacitor-run motor
13. Repulsion type motors
14. That like magnetic fields repel each other
15. Radial
16. A short-circuiting device short circuits the commutator.
17. It operates on the principle of a rotating magnetic field produced in the stator winding.
18. The shading coil located on one side of the pole piece
19. By removing the stator core and turning it around
20. By reversing the current flow through alternate poles
21. Because the stator winding has much higher impedance than most induction motors
22. 3600 RPM
23. 1200 RPM
24. Stepping motors move in steps or increments each time direct current is applied.
25. That like magnetic poles repel and unlike poles attract

26. That there are two windings wound together
27. To permit better stepping resolution
28. Two
29. 72 RPM
30. By applying direct current to the motor windings
31. Because it operates on both direct and alternating current
32. To counteract the inductive reactance of the armature
33. By reversing the armature leads with respect to the series field leads
34. In series with the series field
35. Apply alternating current to the armature and connect a voltmeter across the series field. Set the brushes at the point of lowest induced voltage indicated by the voltmeter.
36. Apply alternating current to the armature and connect a voltmeter to the compensating winding. Set the brushes at the highest induced voltage indicated by the voltmeter.

Tests

ATOMIC STRUCTURE AND OHM'S LAW

NAME _____

Circle the letter beside the correct answer.

1. Which subatomic part of an atom contains a negative charge?
   A. Proton
   B. Electron
   C. Neutron
   D. Nucleus

2. The nucleus of an atom is formed by combining which two subatomic parts?
   A. Proton and electron
   B. Neutron and electron
   C. Proton and neutron
   D. The nucleus is composed of protons, neutrons, and electrons

3. What is electricity?
   A. A flow of electrons
   B. A flow of protons
   C. A flow of neutrons
   D. No one really knows

4. What are VALENCE electrons?
   A. Special electrons that contain a positive charge
   B. Special electrons that contain a negative charge
   C. Electrons that spin in a counterclockwise direction
   D. Electrons located in the outermost orbit of an atom

5. What is a COULOMB?
   A. A measurement of resistance
   B. The amount of voltage contained in a single-cell battery
   C. A quantity of electrons equal to 6.25×10^{18}
   D. A measurement of electrical power

6. What is a conductor?
   A. A material that will permit electricity to flow through it easily
   B. A material that hinders the flow of electricity
   C. A material designed to heat up as current flows through it
   D. A material designed to produce a light as current flows through it

7. An insulator is made from a material that generally contains:
   A. 1 or 2 valence electrons
   B. 3 or 4 valence electrons
   C. 4 valence electrons
   D. 7 or 8 valence electrons

8. What is an insulator?
   A. A material that will permit electricity to flow through it easily
   B. A material that hinders the flow of electricity
   C. A material designed to heat up as current flows through it
   D. A material designed to produce a light as current flows through it

9. A semiconductor is a material that generally contains:
   A. 1 or 2 valence electrons
   B. 3 or 4 valence electrons
   C. 4 valence electrons
   D. 7 or 8 valence electrons

10. What is direct current?
    A. The current that is applied directly to a particular component in a circuit
    B. Current that flows in only one direction
    C. Current that flows first in one direction and then in the other
    D. Current that can be obtained only by having two HOTs and a NEUTRAL

11. Which definition below best defines voltage?
    A. Electrical pressure
    B. The difference in potential between two points
    C. The force that drives or pushes the electricity through the wires
    D. All of the above

12. Which electrical quantity is used to measure the amount of power in a circuit?
    A. Volts
    B. Amps
    C. Ohms
    D. Watts

13. An electric iron has a current draw of 12.5A when connected to a 120-V source. What is the resistance of the heating element in the iron?
A. 0.104Ω
B. 9.6 Ω
C. 1500 Ω
D. 1.302 Ω

14. How much power is being used by the iron in question 13?
A. 1152 W
B. 1.302 W
C. 1500 W
D. 9.6 W

15. A 100-W light bulb is connected to 120 V. How much current will flow through the bulb?
A. 1.2 A
B. 3.2 A
C. 0.5 A
D. 0.8 A

16. An 18-Ω resistor produces 3200 W of heat. How much voltage is connected across the resistor?
A. 240V
B. 177.78 V
C. 120V
D. 180V

17. A resistor produces 1200 W of heat when connected to a 480-V source. What is the resistance of the resistor?
A. 0.4 Ω
B. 2.5 Ω
C. 3000 Ω
D. 192 Ω

18. A 24-Ω resistive heating element produces 1800 W of heat. How much current flows through the element?
A. 75 A
B. 43,200 A
C. 8.66 A
D. 0.876 A

19. 18 A of current flows through an 8-Ω resistor. How much power is being used?
A. 2592 W
B. 144 W
C. 2.25 W
D. 1875W

20. How much voltage is connected across the resistor in question 19?
A. 120 V
B. 72 V
C. 144 V
D. 240 V

MAGNETISM AND RESISTORS

NAME _____

1. What are lodestones?

   _____

2. What are the three basic classifications of magnetic materials?

   A. _____

   B. _____

   C. _____

3. Define the following terms:

   Permeability _____

   _____

   Flux density _____

   _____

   Re1uctance _____

   _____

   Saturation _____

   _____

   Residual magnetism _____

   _____

   Coercive force _____

   _____

   Retentivity _____

   _____

4. A coil of wire produces 100 flux lines per square inch when current flows through it. When a metal core is inserted in the coil, the flux lines increase to 100,000 lines per square inch. What is the permeability of the metal core?

   _____

5. According to the laws of magnetism, do like poles of a magnet attract or repel?

   _____

6. What is the strength of an electromagnet proportional to?

   _____

7. If the direction of current flow through an electromagnet is reversed, will the polarity of the magnetic field reverse or remain the same?

   _____

8. Explain how to demagnetize an object.

   _____

   _____

   _____

   _____

9. What is the advantage of a metal film resistor over a composition carbon resistor?

   _____

10. What is a variable resistor?

   _____

11. What are the two primary functions a resistor is used for in an electric circuit?

   A. _____

   B. _____

12. What is a potentiometer?

   _____

13. Fill in the chart below.

First Band	Second Band	Third Band	Fourth Band	Value
RED	VIOLET	RED	GOLD	
				51000 10%
YELLOW	ORANGE	BROWN	SILVER	
				120000 5%
BROWN	BLACK	BLACK	RED	
				750 20%

14. Many wire wound resistors are wound on a ceramic core material that is hollow in the center. When these resistors are mounted, should they be mounted vertically or horizontally?

_____

15. A metal film resistor is marked with the following color bands: brown, black, black, brown, brown. What is the ohmic value of this resistor, and what is its tolerance?

Ohmic value _____ Tolerance_____

SERIES, PARALLEL, AND COMBINATION CIRCUITS

Name_____

Circle the letter beside the correct answer.

1. Three resistors having a value of 10 Ω each are connected in series to a 300-V power source. What is the total resistance of this circuit?
   A. 10 Ω
   B. 30 Ω
   C. 15 Ω
   D. 3 Ω

2. How much current will flow in the circuit described in question 1?
   A. 10A
   B. 30A
   C. 3A
   D. 300A

3. Assume that three resistors of equal value are connected in series. If the first resistor has a current flow of 5 A through it, what is the total amount of current flow in the circuit?
   A. 15 A
   B. 10 A
   C. 5 A
   D. The correct answer cannot be determined without knowing the amount of voltage applied to the circuit.

4. Three resistors are connected in series. Resistor R_1 has a value of 60 Ω, resistor R_2 has a value of 40 Ω, and resistor R_3 has a value of 50 Ω. A voltage drop of 30 V is measured across resistor R_1. What is the voltage dropped across resistor R_3?
   A. 75 V
   B. 20 V
   C. 25 V
   D. 30 V

5. What is the total amount of voltage connected to the circuit described in question 4?
   A. 75 V
   B. 20 V
   C. 25 V
   D. 30 V

6. Three resistors having a value of 30 Ω each are connected in parallel to a 300-V power source. What is the total resistance of this circuit?
   A. 30 Ω
   B. 10 Ω
   C. 90 Ω
   D. 300 Ω

7. What will be the total amount of current flow in the circuit described in question 6?
   A. 30 A
   B. 10 A
   C. 3 A
   D. 9 A

8. A circuit contains three resistors connected in parallel. Resistor R_1 has a value of 18 Ω, resistor R_2 has a value of 25 Ω, and resistor R_3 has a value of 30 Ω. Resistor R_2 has a current flow of 0.25A through it. How much current flows through resistor R_3?
   A. 0.25 A
   B. 0.166 A
   C. 0.278 A
   D. 0.208 A

9. What is the total amount of current flow in the circuit described in question 8?
   A. 0.25 A
   B. 0.166 A
   C. 0.805 A
   D. 0.644 A

To answer the following questions, refer to Figure 1.

10. What is the total resistance of the circuit shown in Figure 1?

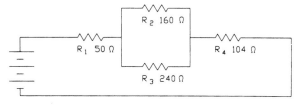

FIGURE 1 Series, parallel, combination circuit test

   A. 250 Ω
   B. 554 Ω
   C. 24.98 Ω
   D. 129.77 Ω

11. Assume that resistor R_2 has a current flow of 0.25 A. What is the voltage drop across resistor R_3?
    A. 40 V
    B. 36 V
    C. 72 V
    D. 52.6 V

12. Assume that resistor R_3 has a current flow of 0.125 A. How much current will flow through resistor R_4?
    A. 0.125 A
    B. 0.1875 A
    C. 0.3125 A
    D. 0.288 A

13. Assume that a voltage of 15 V is dropped across resistor R_1. How much voltage is dropped across resistor R_4?
    A. 15 V
    B. 31.2 V
    C. 62.5 V
    D. 12.8 V

14. If resistor R_1 has a voltage drop of 15 V, how much voltage is dropped across resistor R_2?
    A. 48 V
    B. 72 V
    C. 31.2 V
    D. 28.8 V

15. Assume that resistor R_4 has a current flow of 0.06 A. What is the total amount of current flow in this circuit?
    A. 0.24 A
    B. 0.12 A
    C. 0.06 A
    D. 1.2A

MEASURING INSTRUMENTS

NAME _____

Circle the letter beside the correct answer.

1. What is the principle of operation of a d'Arsonval meter movement?
   A. The attraction of two magnetic fields
   B. The fact that like magnetic fields repel
   C. Induced voltages produce eddy currents that repel each other
   D. Current flowing through a coil produces a change in temperature, which causes the pointer to turn

2. The d'Arsonval meter movement:
   A. Operates on both AC and DC current
   B. Operates on AC current only
   C. Operates on DC current only
   D. Depends on induced voltage and current for operation

3. Which of the following is true concerning voltmeters?
   A. Voltmeters have a very low resistance and must be connected in series with a load.
   B. Voltmeters have a very high resistance and must be connected in series with a load.
   C. Voltmeters have a very low resistance and must be connected in parallel with the device being tested.
   D. Voltmeters have a very high resistance and must be connected in parallel with the device being tested.

4. Which of the following is true concerning ammeters?
   A. Ammeters have a very low resistance and must be connected in series with a load.
   B. Ammeters have a very high resistance and must be connected in series with a load.
   C. Ammeters have a very low resistance and must be connected in parallel with the device being tested.
   D. Ammeters have a very high resistance and must be connected in parallel with the device being tested.

5. DC current is often measured with a shunt by connecting the shunt in series with the load and measuring the voltage drop across the shunt. What is the standard voltage value of a DC ammeter shunt?
   A. 1V
   B. 5 mV
   C. 50 mV
   D. 2.5 V

6. What is the standard output current for a current transformer when it is connected to an AC ammeter?
   A. 1 A
   B. 5 A
   C. 50 mA
   D. 1 mA

7. What electrical value is measured by an ohmmeter?
   A. Volts
   B. Current
   C. Power
   D. Resistance

8. What is the input impedance of most digital multi-meters?
   A. 20,000 ohms per volt on most ranges
   B. 5000 ohms per volt on most ranges
   C. 1,000,000 ohms on all ranges
   D. 10,000,000 ohms on all ranges

9. What electrical quantity is measured by an oscilloscope?
   A. Volts
   B. Current
   C. Power
   D. Resistance

10. An oscilloscope shows that a wave form completes one complete cycle every 28 microseconds. What is the frequency of the wave form?
   A. 28,000,000 Hz
   B. 60 Hz
   C. 35,714.3 Hz
   D. 114,238.6 Hz

11. What type of meter must be used to measure the amount of power used in a circuit?
    A. Ammeter
    B. Wattmeter
    C. Ohmmeter
    D. Voltmeter

12. A clamp-on ammeter is capable of measuring both DC and AC currents. What is the principle of operation for this type of clamp-on ammeter?
    A. A voltage divider network of precision resistors
    B. A series of precision shunts
    C. A current transformer
    D. A Hall effect generator

13. A DC voltmeter has a resistance of 20,000 ohms per volt. What is the resistance of the meter if the selector is set for a full-scale value of 300 V?
    A. 20,000 Ω
    B. 300 Ω
    C. 6,000,000 Ω
    D. 20,000,000 Ω

14. What is measured on the X axis of an oscilloscope?
    A. Time
    B. Resistance
    C. Volts
    D. Current

15. A shunt is used to measure DC current. When a current of 10 A flows through the shunt, a voltage of 50 mV (0.050 V) is dropped across the shunt. What is the resistance of the shunt?
    A. 0.005 Ω
    B. 0.5 Ω
    C. 200 Ω
    D. 5 Ω

16. A d'Arsonval meter movement has a resistance of 1,000 Ω and a full scale current of 1 mA (0.001 A). How much resistance should be connected in series with the meter to permit it to indicate a full-scale voltage of 25 V?
    A. 25,000 Ω
    B. 24,000 Ω
    C. 250 Ω
    D. 240 Ω

17. Which of the meters should never be connected to alive circuit?
    A. Ammeter
    B. Voltmeter
    C. Oscilloscope
    D. Ohmmeter

18. Which of the meter movements listed is not considered to be a moving-iron type of meter?
    A. Radial vane
    B. Concentric vane
    C. Moving coil
    D. Plunger

19. A meter movement has a full-scale voltage of 0.1 V. The meter is to be used as a DC ammeter with a full-scale value of 50 A. What should be the resistance of the shunt connected in series with the load?
    A. 5 Ω
    B. 500 Ω
    C. 0.002 Ω
    D. 0.005 Ω

20. A voltmeter has a resistance of 5000 ohms per volt. The meter is set on a 600-V scale. What is the resistance of the meter?
    A. 5000 Ω
    B. 3,000,000 Ω
    C. 0.12 Ω
    D. 8.333 Ω

CONDUCTORS

NAME _____

NOTE: The student should be permitted to use the National Electrical Code book and the wire tables located in the appendix.

Refer to the National Electrical Code to answer the following questions:

1. A copper conductor with type THHN insulation must carry 166 A. The wire is to be used in a raceway with two other conductors of the same size. What size conductor should be used?
   A. 4/0 AWG
   B. 1/0 AWG
   C. 3/0 AWG
   D. 2/0 AWG

2. A 500-kCM copper conductor with type THWN insulation is used in an area that has an ambient temperature of 53°C. What is the maximum current-carrying capacity of this conductor?
   A. 380 A
   B. 254.6 A
   C. 131.2 A
   D. 307.1 A

3. What is the maximum temperature rating of type XHHW insulation when used in a wet location?
   A. 90°C
   B. 194°F
   C. 185°F
   D. 75°C

4. A #10 AWG copper conductor with type THWN insulation is to be run in free air. What is the maximum ampacity of this conductor if the ambient air temperature is 40°C?
   A. 40 A
   B. 50 A
   C. 35.2 A
   D. 44 A

5. Six #1/0 AWG aluminum conductors are to be run in a single conduit. Each conductor has type THWN insulation, and the ambient air temperature is 30°C. How much current is each conductor permitted to carry?
   A. 150 A
   B. 125 A
   C. 96 A
   D. 120 A

6. Nine #12 AWG copper conductors are run in a conduit. Each conductor has THHN insulation, and the ambient air temperature is 118°F. How much current is each conductor permitted to carry?
   A. 17.2 A
   B. 30 A
   C. 24.6 A
   D. 21 A

Refer to the American Wire Gauge Table in the appendix to answer the following questions:

Formulas and Chart

$$R = \frac{K \times L}{CM} \quad L = \frac{R \times CM}{K} \quad CM = \frac{K \times L}{R}$$

$$K = \frac{CM \times R}{L}$$

RESISTIVITY CF MATERIALS (K)

MATERIAL	Ω per Mil-Foot at 20°C	Temp. Coeff. Ω Per °C
Aluminum	17	0.004
Carbon	22,000 Aprx.	-0.0004
Constantan	295	0.000002
Copper	10.4	0.0039
Gold	14	0.004
Iron	60	0.0055
Lead	126	0.0043
Manganin	265	0.000000
Mercury	590	0.00088
Nichrome	675	0.0002
Nickel	52	0.005
Platinum	66	0.0036
Silver	9.6	0.0038
Tungsten	33.8	0.005

7. Name four factors that determine the resistance of a wire.

A. _____

B. _____

C. _____

D. _____

8. A copper #24 AWG copper wire is 125 feet long. What is the resistance of this wire? Assume an ambient temperature of 20°C
   A. 0.027 Ω
   B. 1.16 Ω
   C. 3.22 Ω
   D. 4.06 Ω

9. A copper conductor must be run a distance of 250 feet. The wire cannot have a resistance greater than 0.16 Ω. What is the smallest AWG size of copper conductor that can be used in this installation?
   A. #2 AWG
   B. #4 AWG
   C. #6 AWG
   D. #8 AWG

10. A piece of #16 AWG wire is 50 feet long and has a resistance of 1.162 Ω. Of what material is this wire made?
    A. Iron
    B. Aluminum
    C. Copper
    D. Nichrome

BATTERIES

NAME _____

1. What determines the amount of voltage produced by a voltaic cell?

_____

_____

2. What determines the current capacity of a cell?

_____

_____

3. Explain the difference between a primary cell and a secondary cell.

Primary _____

Secondary_____

4. What voltage is produced by each of the cells listed below?

Carbon-Zinc _____ Alkaline _____

Ni-cad _____ Lead Acid _____

5. What is the electrolyte used in a lead-acid automobile battery?

_____

6. What is the specific gravity of the electrolyte used in a lead-acid battery?

_____

7. What type of gas is given off when a lead-acid battery is being discharged?

_____

8. When a lead-acid battery is discharged, a layer of substance builds up on the battery plates. When the battery is charged properly, this substance dissolves and part of it recombines with the water to form new acid. What is this substance that forms on the plates? _____

9. A 12-V 80 A-hr lead-acid battery is to be charged. The charging current should no be greater than _____ A.

10. Four lead-acid batteries, each rated at 12 V and 40 A-hr are connected in series. What is the total voltage and amp hour capacity of this connection?

Volts _____

A-hr _____

11. If the four batteries in the above problem were connected in parallel, what would be the voltage and amp-hour capacity of the connection?

Volts _____

A-hr _____

12. A lead-acid battery rated at 12 V and 80 A-hr is to be load tested. A current of _____ A should be used as the test current, and the voltage should not drop below _____V for a period of _____seconds.

13. What device is used to measure the specific gravity of the electrolyte in a cell?

_____

14. What is the specific gravity a measure of?

_____

_____

15. What determines the amount of voltage produced by a solar cell?

_____

16. What determines the amount of current that can be supplied by a solar cell?

_____

17. What is the open circuit voltage of a silicon solar cell in direct sunlight?

_____ V

18. What two factors determine the amount of voltage that can be produced by a thermocouple?

A. _____

B. _____

19. What is a thermopile?

_____

20. What is the piezoelectric effect?

   _____

   _____

BASIC ALTERNATING CURRENT

NAME _____

Circle the letter beside the correct answer.

1. When AC voltage is being measured with a voltmeter, which value of AC voltage will be indicated?
   A. Peak
   B. Average
   C. RMS
   D. Peak-to-peak

2. How many degrees are there in one complete cycle of alternating voltage?
   A. 90
   B. 180
   C. 270
   D. 360

3. A sine wave has a peak or maximum value of 150 V. What is the voltage at 40° on the wave form?
   A. 3.75
   B. 96.4
   C. 75
   D. 60

4. A sine wave has a peak value of 200 V. What is the RMS value of voltage?
   A. 282.9
   B. 400
   C. 141.4
   D. 127.4

5. A sine wave has a peak value of 280 V. What is the average value of voltage?
   A. 178.4
   B. 198
   C. 395.9
   D. 89.2

6. A sine wave has an average value of 68 V. What is the RMS value of voltage?
   A. 96.2
   B. 75.5
   C. 48.1
   D. 43.3

7. A sine wave has an RMS value of 120 V. What is the peak value of voltage?
   A. 84.8
   B. 188.4
   C. 169.7
   D. 76.4

8. What is the relationship of current and voltage in a pure resistive circuit?
   A. The current lags the voltage by 90°.
   B. The current leads the voltage by 90°.
   C. The current and voltage are 180° out of phase with each other.
   D. The current and voltage are in phase with each other.

9. Which description below best describes RMS voltage?
   A. The maximum amount of voltage reached by a wave form
   B. The amount of voltage that would be measured by a DC voltmeter after a sine wave AC voltage had been rectified with a full-wave rectifier.
   C. The amount of voltage that would be measured by a DC voltmeter after a sine wave AC voltage had been rectified with a halfwave rectifier.
   D. The amount of AC voltage that will produce as much power as a like amount of pure DC voltage.

10. Which description below best describes AVERAGE voltage?
   A. The maximum amount of voltage reached by a wave form.
   B. The amount of voltage that would be measured by a DC voltmeter after a sine wave AC voltage had been rectified with a full-wave rectifier.
   C. The amount of voltage that would be measured by a DC voltmeter after a sine wave AC voltage had been rectified with a half-wave rectifier.
   D. The amount of AC voltage that will produce as much power as a like amount of pure DC voltage.

ALTERNATING CURRENT CIRCUITS CONTAINING INDUCTANCE

NAME_____

Circle the letter beside the correct answer.

1. What is the relationship of current and voltage in a pure inductive circuit?
   A. The current lags the voltage by 90°.
   B. The current leads the voltage by 90°.
   C. The current and voltage are 180° out of phase with each other.
   D. The current and voltage are in phase with each other.

2. A 0.25-H inductor is connected to a 120-V, 60-Hz line. What is the inductive reactance of this circuit?
   A. 153.6 Ω
   B. 43.7 Ω
   C. 286.6 Ω
   D. 94.2 Ω

3. A 120-V, 60-Hz AC circuit is connected to an inductor. The current flow in the circuit is 0.35 A. What is the inductive reactance of the circuit?
   A. 131.9 Ω
   B. 342.8 Ω
   C. 42 Ω
   D. 288.9 Ω

4. An inductor with a inductance of 0.06 H is connected to a 480-V, 400-Hz alternator. How much current will flow in this circuit?
   A. 3.18 A
   B. 150.8 A
   C. 21.2 A
   D. 0.85 A

5. Three inductors are connected in series. Their values of inductance are as follows: L_1 = 0.5 H, L_2 = 0.4 H, and L_3 = 0.8 H. If these inductors are connected to a 60-Hz line, what will be the total inductive reactance of the circuit?
   A. 65.56 Ω
   B. 253.4 Ω
   C. 640.9 Ω
   D. 1256 Ω

6. Three inductors connected in series have inductances of 0.012 H, 0.036 H, and 0.024 H. If these inductances are connected to a 277-V, 400-Hz line, how much current will flow in the line?
   A. 16.84 A
   B. 27.14 A
   C. 15.34 A
   D. 1.53 A

7. Three inductors connected in parallel have inductive reactances of 24 Ω, 48 Ω, and 36 Ω. If the inductors are connected to a 60-Hz line, what is the total inductance of the circuit?
   A. 0.0294 H
   B. 0.346 H
   C. 0.0579 H
   D. 0.642 H

8. Three inductors connected in parallel have inductances of 0.75 H, 0.50 H, and 0.60 H. If these inductors are connected to a 400-Hz line, what will be the total inductive reactance of the circuit?
   A. 346.7 Ω
   B. 980.3 Ω
   C. 502.6 Ω
   D. 118.8 Ω

9. An inductor with an inductive reactance of 48 Ω is connected in series with a resistor with a value of 32 Ω. What is the total impedance of this circuit?
   A. 80 Ω
   B. 19.2 Ω
   C. 26.6 Ω
   D. 57.7 Ω

10. An inductor with an inductive reactance of 48 Ω is connected in parallel with a resistor with a value of 32 Ω. What is the total impedance of this circuit?
    A. 80a
    B. 19.2 Ω
    C. 26.6 Ω
    D. 57.7 Ω

11. A circuit contains resistance and inductance connected in series with each other. The resistor has a true power of 560 W. The inductor has a reactive power of 430 VARs. What is the power factor of this circuit?
    A. 79.3%
    B. 76.8%
    C. 60.9%
    D. 83%

12. A circuit contains resistance and inductance connected in parallel with each other. There is an apparent power of 48,000 VA and a true power of 34,000 W. How much reactive power is in this circuit?
    A. 19,902.4 VARs
    B. 58,851.7 VARs
    C. 33,882.1 VARs
    D. 27,744.8 VARs

13. An RI series circuit contains a power factor of 76.4%. How many degrees out of phase are the current and voltage with each other?
    A. 40.2°
    B. 53.13°
    C. 76.6°
    D. 0.235°

14. What electrical quantity is used to measure INDUCTANCE?
    A. Farad
    B. Coulomb
    C. Hertz
    D. Henry

15. What would limit the amount of current flow in a pure inductive circuit?
    A. Resistance of the wire
    B. Inductive reactance
    C. Impedance
    D. Capacitive reactance

ALTERNATING CURRENT CIRCUITS CONTAINING CAPACITANCE

NAME_____

1. List three factors that determine the amount of capacitance a capacitor will have.

   A. _____

   B. _____

   C. _____

2. What unit of measure is used for measuring capacitance?
   A. Henry
   B. Hertz
   C. Farad
   D. Ohm

3. A 25-µF capacitor is connected to a 480-V 60-Hz line. How much current will flow in this circuit?
   A. 19.2 A
   B. 4.52 A
   C. 0.052 A
   D. 1.06 A

4. A 5-µF capacitor is connected to a 400-Hz line. What is the capacitive reactance of this circuit?
   A. 159.15 Ω
   B. 235.7 Ω
   C. 53.5 Ω
   D. 79.6 Ω

5. A capacitor connected to a 240-V, 60-Hz circuit has a current flow of 4.5 A. If the capacitor is connected to a 240-V, 400-Hz line how much current will flow?
   A. 30 A
   B. 15 A
   C. 0.96 A
   D. 42 A

6. Three capacitors having values of 30 µF, 45 µF, and 60 µF are connected in series. What is the total capacitance of this circuit?
   A. 135 µF
   B. 43.6 µF
   C. 156 µF
   D. 13.8 µF

7. If the capacitors in question 6 were to be reconnected in parallel, what would be the total capacitance of the circuit?
   A. 135 µF
   B. 43.6 µF
   C. 156 µF
   D. 13.8 µF

8. A resistor and capacitor are connected in series with each other. The resistor has a resistance of 24 Ω, and the capacitor has a capacitive reactance of 16 Ω. What is the total impedance of the circuit?
   A. 36.4 Ω
   B. 28.8 Ω
   C. 40 Ω
   D. 122.7 Ω

9. What is the power factor of the circuit in question 8?
   A. 12%
   B. 53.1%
   C. 83.3%
   D. 76.5%

10. A resistor and capacitor are connected in series. The circuit has a total applied voltage of 120 V. There is a voltage drop of 96 V across the resistor. How much voltage is being dropped across the capacitor?
    A. 24 V
    B. 86 V
    C. 72 V
    D. 38 V

11. A resistor and capacitor are connected in parallel to a 60-Hz line. The resistor has a resistance of 64 Ω, and the capacitor has a capacitive reactance of 56 Ω. What is the total impedance of this circuit?
    A. 32.7 Ω
    B. 85 Ω
    C. 120 Ω
    D. 42.1 Ω

12. A resistor and capacitor are connected in parallel. The resistor has a current flow of 1.8 A through it, and the capacitor has a current flow of 2.2 A through it. What is the total circuit current?
    A. 2.8 A
    B. 0.4 A
    C. 4 A
    D. 0.72 A

13. How many degrees out of phase are the voltage and current with each other in question 12?
    A. 36.6°
    B. 50°
    C. 78°
    D. 18.4°

14. A resistor and capacitor are connected in parallel to a 60-Hz line. The total impedance of the circuit is 8.45 Ω. The resistor has a resistance of 16 Ω. What is the capacitance of the capacitor?
    A. 9.95 µF
    B. 266.6 µF
    C. 38.9 µF
    D. 452.8 µF

15. What is the power factor of the circuit in question 14?
    A. 1.89%
    B. 189.3%
    C. 52.8%
    D. 38.6%

THREE-PHASE CIRCUITS

NAME_____

Circle the letter beside the correct answer.

1. How many degrees out of phase are the voltages of a three-phase system?
   A. 60°
   B. 90°
   C. 120°
   D. 180°

2. Which of the following statements is true for a wye-connected three-phase system?
   A. Line current is higher than phase current by a factor of 1.732.
   B. Phase current is higher than line current by a factor of 1.732.
   C. Line current is less than phase current by a factor of 1.732.
   D. Line current and phase current are the same.

3. Which of the following statements is true for a delta-connected three-phase system?
   A. Line current is higher than phase current by a factor of 1.732.
   B. Phase current is higher than line current by a factor of 1.732.
   C. Line current is less than phase current by a factor of 1.732.
   D. Line current and phase current are the same.

4. Which of the following statements is true for a wye-connected three-phase system?
   A. Line voltage is higher than phase voltage by a factor of 1.732.
   B. Phase voltage is higher than line voltage by a factor of 1.732.
   C. Line voltage is less than phase voltage by a factor of 1.732.
   D. Line voltage and phase voltage are the same.

5. Which of the following statements is true for a delta-connected three-phase system?
   A. Line voltage is higher than phase voltage by a factor of 1.732.
   B. Phase voltage is higher than line voltage by a factor of 1.732.
   C. Line voltage is less than phase voltage by a factor of 1.732.
   D. Line voltage and phase voltage are the same.

6. The windings of a three-phase alternator are connected delta. The alternator is producing a line-to-line voltage of 480 V. What is the voltage across the phase windings?
   A. 480 V
   B. 120 V
   C. 277 V
   D. 831 V

7. The alternator in question 6 is connected to a load that draws a current of 25.98 A on the lines. How much current is flowing in the phase windings?
   A. 25.98 A
   B. 45 A
   C. 15 A
   D. 17.32 A

8. A three-phase wye-connected load is connected to a 480-V three-phase line. The line current to the load is 36 A. If this load is pure resistive, what is the total amount of power being used in this circuit?
   A. 51,840 W
   B. 29,929 W
   C. 7482 W
   D. 12,960 W

9. A delta-connected three-phase load is connected to a 560-V line. The loads have an impedance of 14 Ω per phase. What is the apparent power of this connection?
   A. 22,393 VA
   B. 116,390 VA
   C. 38,797 VA
   D. 67,200 VA

10. If the loads in question 9 were reconnected to form a wye connection instead of a delta, what would be the apparent power of the circuit?
    A. 22,393 VA
    B. 116,390 VA
    C. 38,797 VA
    D. 67,200 VA

SINGLE-PHASE TRANSFORMERS

NAME _____

Circle the letter beside the correct answer.

1. What electrical quantity cannot be changed by a transformer?
   A. Voltage
   B. Current
   C. Impedance
   D. Frequency

2. An isolation transformer:
   A. Has its primary and secondary windings electrically and mechanically separated from each other.
   B. Uses the same winding for both primary and secondary windings.
   C. Produces a lower secondary voltage than primary voltage.
   D. Produces a higher secondary voltage than primary voltage.

3. An autotransformer:
   A. Has its primary and secondary windings electrically and mechanically separated from each other.
   B. Uses the same winding for both primary and secondary windings.
   C. Produces a lower secondary voltage than primary voltage.
   D. Produces a higher secondary voltage than primary voltage.

4. All values of a transformer are proportional to:
   A. The type of core material used to construct the transformer.
   B. The number of turns of wire on the primary winding.
   C. The ratio of primary turns of wire as compared with secondary turns of wire.
   D. The impedance of the secondary winding.

5. What is self-induction?
   A. The action that takes place when a coil induces a voltage into the core material
   B. The action that takes place when a coil induces a voltage into itself
   C. The action that takes place when a coil induces a voltage into the surrounding air
   D. The action that takes place when a coil induces a voltage into another coil

6. What is mutual induction?
   A. The action that takes place when a coil induces a voltage into the core material
   B. The action that takes place when a coil induces a voltage into itself
   C. The action that takes place when a coil induces a voltage into the surrounding air
   D. The action that takes place when a coil induces a voltage into another coil

7. A transformer has a turns ratio of 2.5:1. The primary voltage is 208 V. What is the voltage of the secondary winding?
   A. 520 V
   B. 360.256 V
   C. 83.2 V
   D. 120 V

8. The transformer in question 7 has a secondary current of 18 A. What is the primary current?
   A. 7.2A
   B. 83.2 A
   C. 45 A
   D. 11.555 A

9. A transformer has a turns ratio of 1:3. The primary voltage is 120 V. What is the secondary voltage?
   A. 40 V
   B. 84 V
   C. 240 V
   D. 360 V

10. The transformer in question 9 has a secondary current of 1.2 A. What is the primary current?
    A. 0.4 A
    B. 0.84 A
    C. 2.4 A
    D. 3.6 A

11. A transformer has a primary voltage of 240 V and an apparent power of 360 VA. The secondary voltage is 48 V. What is the secondary current?
    A. 7.5 A
    B. 5 A
    C. 1.5 A
    D. There is not enough information given to answer this question.

12. Fill in the missing values on the figure below.

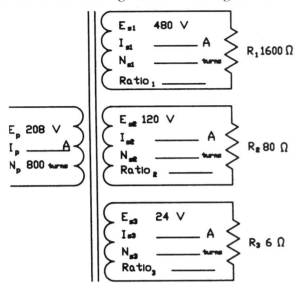

13. On the figure below, fill in the amount of voltage produced across: A-B, A-C, A-D, B-C, B-D, C-D.

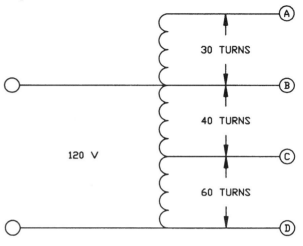

THREE-PHASE TRANSFORMERS

NAME _____

1. Fill in the missing values for the figure below.

Ratio: _____

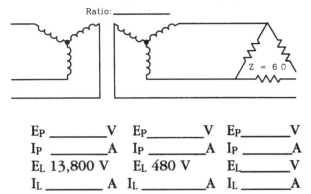

E_P _____V	E_P_____V	E_P_____V
I_P _____A	I_P _____A	I_P _____A
E_L 13,800 V	E_L 480 V	E_L_____V
I_L _____A	I_L _____A	I_L _____A

2. Fill in the missing values for the figure below.

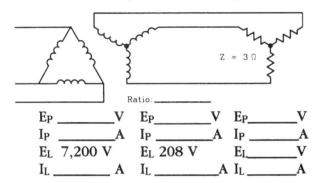

Ratio: _____

E_P _____V	E_P_____V	E_P_____V
I_P _____A	I_P _____A	I_P _____A
E_L 7,200 V	E_L 208 V	E_L_____V
I_L _____A	I_L _____A	I_L _____A

3. Fill in the missing values for the figure below.

Ratio: _____

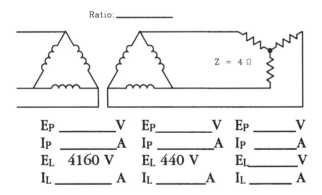

E_P _____V	E_P_____V	E_P _____V
I_P _____A	I_P _____A	I_P _____A
E_L 4160 V	E_L 440 V	E_L_____V
I_L _____A	I_L _____A	I_L _____A

4. Fill in the missing values for the figure below.

Ratio: _____

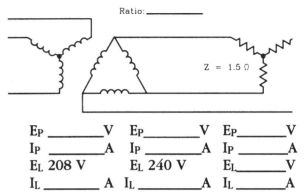

E_P _____V	E_P_____V	E_P_____V
I_P _____A	I_P _____A	I_P _____A
E_L 208 V	E_L 240 V	E_L_____V
I_L _____A	I_L _____A	I_L _____A

5. Connect the three single-phase transformers shown below to form a three-phase bank with a wye primary and a delta secondary. Connect the primary windings to the incoming power lines and the secondary windings to the load lines.

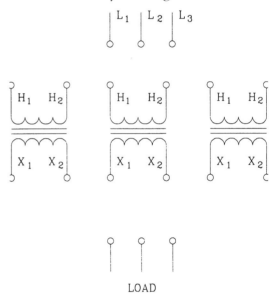

6. Three single-phase transformers have a rating of 60 KVA each. If these transformers are connected to form a three-phase bank, what will be the KVA rating of the bank?

7. Two single-phase transformers having a capacity of 50 KVA each are connected to form an open delta connection. What is the total KVA capacity of this connection?

DIRECT CURRENT MACHINES

NAME_____

Circle the letter beside the correct answer.

1. The operating characteristic of a series generator is:
   A. The output voltage decreases as load increases.
   B. The output voltage is the same at no load as it is at full load.
   C. The output voltage increases as load increases.
   D. Load current has no effect on the output voltage.

2. The operating characteristic of a shunt generator is:
   A. The output voltage decreases as load increases.
   B. The output voltage is the same at no load as it is at full load.
   C. The output voltage decreases as load increases.
   D. Load current has no effect on the output voltage.

3. Which of the following best describes a series field?
   A. Wound with a few turns of large wire and connected in parallel with the armature
   B. Wound with many turns of small wire and connected in parallel with the armature
   C. Wound with many turns of small wire and connected in series with the armature
   D. Wound with a few turns of large wire and connected in series with the armature

4. Which of the following best describes a shunt field?
   A. Wound with a few turns of large wire and connected in parallel with the armature
   B. Wound with many turns of small wire and connected in parallel with the armature
   C. Wound with many turns of small wire and connected in series with the armature
   D. Wound with a few turns of large wire and connected in series with the armature

5. When a DC machine is cumulative-compounded:
   A. The series and shunt fields aid each other in the production of magnetism.
   B. The series and shunt fields are connected in series with each other.
   C. The series and shunt fields oppose each other in the production of magnetism.
   D. The series and shunt fields are connected in parallel with each other.

6. What is the job of the commutator in a DC generator?
   A. It changes the AC voltage produced in the armature into DC voltage.
   B. It provides a connection to the outside circuit via the brushes.
   C. It provides a connection for the windings in the armature.
   D. All of the above.

7. What type of armature winding is used for machines designed for use on low-voltage, high-current circuits?
   A. Wave winding
   B. Frogleg winding
   C. Lap winding
   D. Compound winding

8. The twisting of the main magnetic field is called:
   A. Commutation
   B. Armature reaction
   C. Reluctance
   D. Regulation

9. What are interpoles?
   A. Windings on the main pole pieces connected in series with the armature
   B. Small pole pieces located between the main field poles and connected in series with the armature
   C. Small pole pieces located between the main field poles and connected in parallel with the armature
   D. Windings on the main pole pieces connected in parallel with the armature

10. What determines the voltage regulation of a DC generator?
    A. The resistance of the armature
    B. The resistance of the shunt field
    C. The resistance of the series field
    D. The number of turns of wire in the armature

11. What determines the output voltage of a DC generator?
    A. The number of turns of wire in the armature
    B. The strength of the magnetic field of the pole pieces
    C. The speed of the armature
    D. All of the above

12. A DC generator is over-compounded when:
    A. The output voltage is less at full load than it is at no load.
    B. The output voltage is the same at full load as it is at no load.
    C. The output voltage is higher at full load than it is at no load.
    D. The load current has no effect on output voltage.

13. What type of armature winding is used for machines intended for use in high-voltage, low-current circuits?
    A. Wave winding
    B. Frogleg winding
    C. Lap winding
    D. Compound winding

14. A generator is self-excited when:
    A. An outside source of alternating current is used to produce magnetism in the pole pieces.
    B. An initial voltage is produced by residual magnetism in the pole pieces. This voltage produces a current flow through the shunt field, which causes an increase in the output voltage.
    C. An outside source of direct current is used to produce magnetism in the pole pieces.
    D. An initial voltage is produced in the series field, which produces an increase in the output voltage of the generator.

15. When a generator produces current to supply power to a load, the armature becomes hard to turn. This increase of required turning force is called countertorque. What two factors determine the amount of countertorque produced by the generator?
    A. The horsepower of the prime mover and the connected load
    B. The strength of the magnetic field of the series field and the strength of the shunt field
    C. The strength of the magnetic field of the pole pieces and the strength of the magnetic field of the armature
    D. The resistance of the armature and the resistance of the shunt field

16. The leads of a DC machine are brought out to the terminal connection box. What markings are used to identify the shunt field leads?
    A. F_1 and F_2
    B. A_1 and A_2
    C. S_1 and S_2
    D. C_1 and C_2

17. When a DC machine is connected long shunt compound:
    A. The armature is connected in parallel with both the series and shunt fields.
    B. The shunt field is connected in parallel with both the armature and series field.
    C. The shunt field is connected in parallel with the armature and in series with the series field.
    D. The shunt field is connected in series with the armature and in parallel with the series field.

18. What procedure can be used to set the brushes to the neutral plane position in a DC machine?
    A. Connect a DC voltmeter across the armature leads and apply direct current to the shunt field. Move the brushes until the meter indicates 0 V.
    B. Connect an AC voltmeter to the armature and apply alternating current to the shunt field. Move the brushes until the meter indicates 0 V.
    C. Apply direct current to the armature and connect a DC voltmeter across the shunt field leads. Move the brushes until the meter indicates 0 V.
    D. Apply alternating current to the armature and connect an AC voltmeter across the shunt field leads. Move the brushes until the meter indicates 0 V.

19. How is the amount of compounding of a DC generator controlled?
    A. A low-value variable resistor is connected in series with the armature.
    B. A low-value variable resistor is connected in parallel with the series field.
    C. A large-value variable resistor is connected in series with the shunt field.
    D. A large-value variable resistor is connected in parallel with the shunt field.

20. How is the output voltage of a direct current generator that does not contain a separate voltage regulator generally controlled?
    A. A low-value variable resistor is connected in series with the armature.
    B. A low-value variable resistor is connected in parallel with the series field.
    C. A large-value variable resistor is connected in series with the shunt field.
    D. A large-value variable resistor is connected in parallel with the shunt field.

21. What type of DC motor should never be operated under a no-load condition?
    A. Series
    B. Shunt
    C. Cumulative-compound
    D. Differential-compound

22. Why should a DC shunt or compound motor be started with maximum field excitation?
    A. It causes an increase of starting current, which produces higher starting torque.
    B. It reduces the starting torque of the motor and prevents damage to the connected load.
    C. It causes a decrease in the amount of starting current.
    D. A shunt or compound motor should never be started with maximum field excitation.

23. What determines the speed regulation of a DC motor?
    A. Counter-EMF
    B. The number of turns of wire in the armature
    C. The strength of the shunt field
    D. The resistance of the armature winding

24. What is the most accepted method for reversing the direction of rotation for a DC motor?
    A. Reversing the connection to the series field leads
    B. Reversing the connection of the armature leads
    C. Reversing the connection of the shunt field leads
    D. Reversing the polarity of voltage supplied to the motor

25. What two factors determine the amount of torque produced by a DC motor?
    A. Strength of the magnetic field of the armature and strength of the magnetic field of the pole pieces
    B. The number of turns of wire in the shunt field and the number of turns of wire in the armature winding
    C. The strength of the magnetic field of the pole pieces and the speed of the armature
    D. The resistance of the armature winding and the speed of rotation

THREE-PHASE MOTORS

NAME _____

1. What are the three major types of three-phase motors?

   A. _____

   B. _____

   C. _____

2. What is synchronous speed?

   _____

3. What two factors determine the synchronous speed of a three-phase motor?

   A. _____

   B. _____

4. A three-phase squirrel cage motor has a full-load speed of 1175 RPM when connected to a 60-Hz line. How many stator poles per phase does the motor have?

   _____

5. What does the code letter on the name plate of a squirrel cage motor indicate about the motor?

   _____

6. What three things determine the amount of torque produced by an induction motor?

   A. _____

   B. _____

   C. _____

7. A three-phase motor has three sliprings on its rotor shaft. What type of motor is this?

   _____

8. What type of three-phase motor operates at synchronous speed?

   _____

9. Explain why the starting torque of a three-phase wound rotor induction motor is higher per amp than that of a three-phase squirrel cage induction motor.

   _____

   _____

   _____

   _____

10. What are the amortisseur windings in a three-phase synchronous motor?

    _____

    _____

11. What must be done to a synchronous motor to make it have a leading power factor?

    _____

    _____

12. Is the excitation current of a synchronous motor AC or DC?

    _____

13. Is the excitation current of a synchronous motor applied before the motor is started or after the motor is started?

    _____

14. How is the direction of rotation of a three-phase motor reversed?

    _____

15. What type of three-phase motor changes its speed by reconnecting the stator winding to form a different number of poles?

    _____

SINGLE-PHASE MOTORS

NAME_____

1. What are the three major types of single-phase split-phase motors?

   A. _____

   B. _____

   C. _____

2. How is the direction of rotation of a single-phase split-phase motor reversed?

   _____

3. Explain how the capacitor of a capacitor-start induction-run motor increases starting torque.

   _____

   _____

   _____

4. What function does the centrifugal switch in a split-phase motor perform?

   _____

   _____

5. What type of split-phase motor does not generally use a centrifugal switch?

   _____

6. What is the advantage of a repulsion motor over the other types of single-phase motors?

   _____

7. How is the direction of rotation of a repulsion motor reversed?

   _____

   _____

8. What two factors determine the speed of the rotating magnetic field of a single-phase induction motor?

   A. _____

   B. _____

9. What type of single-phase motor will operate on AC or DC voltage?

   _____

10. The nameplate of a single-phase motor states that the full-load speed is 17,000 RPM. What type of motor is this?

   _____

11. Are the field windings of a universal motor connected in series with the armature or in parallel with the armature?

   _____

12. How is the direction of rotation of a universal motor reversed?

   _____

13. Can a universal motor use inductive compensation when connected to DC?

   _____

14. Why are the poles of a shaded-pole induction motor shaded?

   _____

   _____

   _____

15. How is the direction of rotation of a shaded-pole induction motor reversed?

   _____

16. What prevents the windings of a multi-speed fan motor from burning up when the motor is operated at low speed?

   _____

17. What will be the synchronous speed of an eight-pole stepping motor when it is connected to a 60-Hz line?_____ RPM

ANSWERS TO TESTS

ATOMIC STRUCTURE AND OHM'S LAW
Answers

1. B
2. C
3. A
4. D
5. C
6. A
7. D
8. B
9. C
10. B
11. D
12. D
13. B
14. C
15. D
16. A
17. D
18. C
19. A
20. B

MAGNETISM AND RESISTORS
Answers

1. Natural magnets

2. Ferromagnetic, paramagnetic, diamagnetic

3. Permeability: A material's ability or willingness to become magnetized

   Flux density: The number of magnetic flux lines per square inch. The measure of the strength of a magnetic field.

   Reluctance: Resistance to magnetism

Saturation: Th epoint at which a material can hold no more lines of magnetic flux

Residual magnetism: The amount of magnetism left in a material after the magnetizing force has been removed

Coercive force: A material's ability to retain magnetism

Retentivity: A material's ability to retain magnetism

4. 1000 (100,000/100 = 1000)

5. Repel

6. Ampere turns

7. Reverse

8. Place it in an alternating current magnetic field and draw the object away. Heating it. Strike it with a hard object.

9. It does not change its resistance value with age.

10. A resistor whose value can be changed over a certain range

11. A. To limit current flow through a circuit
    B. As a voltage divider

12. A variable resistor used as a voltage divider

13.

First Band	Second Band	Third Band	Fourth Band	Value
RED	VIOLET	RED	GOLD	2700 5%
GREEN	BROWN	ORANGE	SILVER	51000 10%
YELLOW	ORANGE	BROWN	SILVER	430 10%
BROWN	RED	YELLOW	GOLD	120000 5%
BROWN	BLACK	BLACK	RED	10 2%
VIOLET	GREEN	BROWN	NONE	750 20%

14. Vertically

15. 1000 Ω; 1%

SERIES, PARALLEL, AND COMBINATION CIRCUITS
Answers

1. B
2. A
3. C
4. C
5. A
6. B
7. A
8. D
9. C
10. A
11. A
12. C
13. B
14. D
15. C

MEASURING INSTRUMENTS
Answers

1. B
2. C
3. D
4. A
5. C
6. B
7. D
8. D
9. A
10. C
11. B
12. D
13. C
14. A

15. A
16. B
17. D
18. C
19. C
20. B

CONDUCTORS
Answers

1. B
2. B
3. D
4. D
5. C
6. A
7. A. Type of material the wire is made of

   B. Circular mil area of the wire

   C. Length of the wire

   D. Temperature of the wire
8. C
9. D
10. A

BATTERIES
Answers

1. The materials it is made of
2. The area of the plates
3. Primary cells cannot be recharged. Secondary cells can be recharged.
4. Carbon-zinc: 1.5 V Alkaline: 1.5 V Ni-cad: 1.2 V Lead-acid: 2 V
5. Dilute sulfuric acid (H_2SO_4)
6. 1.215 to 1.28
7. Hydrogen
8. Lead sulfate ($PbSO_4$)
9. 8A
10. 48 V and 40 A-hr

11. 12 V and 160 A-hr

12. 240 A 9.6 V 15 seconds

13. Hydrometer

14. The amount of acid in solution as compared with the amount of water

15. The material it is made of

16. The surface area

17. 0.5 V

18. A. The materials it is made of
    B. The difference in temperature between the two junctions

19. A group of thermocouples connected in series

20. The ability of some materials to produce a voltage when subjected to pressure

BASIC ALTERNATING CURRENT

Answers

1. C
2. D
3. B
4. C
5. A
6. B
7. C
8. D
9. D
10. B

ALTERNATING CURRENT CIRCUITS CONTAINING INDUCTANCE

Answers

1. A
2. D
3. B
4. A
5. C
6. D
7. A

8. C
9. D
10. C
11. A
12. C
13. A
14. D
15. B

ALTERNATING CURRENT CIRCUITS CONTAINING CAPACITANCE

Answers

1. A. The area of the plates
   B. The distance between the plates
   C. The type of dielectric used

2. C
3. B
4. D
5. A
6. D
7. A
8. B
9. C
10. C
11. D
12. A
13. B
14. B
15. C

THREE-PHASE CIRCUITS
Answers

1. C
2. D
3. A
4. A
5. D
6. A
7. C
8. B
9. D
10. A

SINGLE-PHASE TRANSFORMERS
Answers

1. D
2. A
3. B
4. C
5. B
6. D
7. C
8. A
9. D
10. D
11. A
12. I_p 2.02 A

   I_{s_1} 0.3 A N_{s_1} 1846 turns $Ratio_1$ 1:2.308

   I_{s_2} 1.5 A N_{s_2} 462 turns $Ratio_2$ 1.733:1

   I_{s_3} 4 A N_{s_3} 92 turns $Ratio_3$ 8.667:1

13. A-B 36V A-C 84 V A-D 156 V B-C 48 V B-D 120 V C-D 72 V

THREE-PHASE TRANSFORMERS
Answers

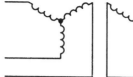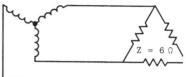

1. Ratio: 28.75:1

E_P 7967.7 V	E_P 277 V	E_P 480 V
I_P 4.819 A	I_P 138.6 A	I_P 80 A
E_L 13,800 V	E_L 480 V	E_L 480 V
I_L 4.819 A	I_L 138.6 A	I_L 138.6 A

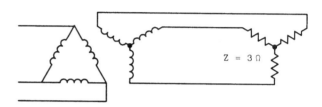

2. Ratio: 60:1

E_P 7200 V	E_P 120 V	E_P 120 V
I_P 0.667 A	I_P 40 A	I_P 40 A
E_L 7200 V	E_L 208 V	E_L 208 V
I_L 1.155 A	I_L 40 A	I_L 40 A

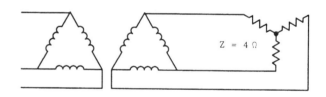

3. Ratio: 9.454:1

E_P 4160 V	E_P 440 V	E_P 254 V
I_P 3.878 A	I_P 36.669 A	I_P 63.51 A
E_L 4160 V	E_L 440 V	E_L 440 V
I_L 6.717 A	I_L 63.51 A	I_L 63.51 A

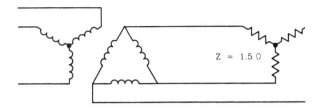

4. Ratio: 1:2

E_P 120 V	E_P 240 V	E_P 138.568 V
I_P 106.673 A	I_P 53.336 A	I_P 92.389 A
E_L 208 V	E_L 240 V	E_L 240 V
I_L 106.673 A	I_L 92.389 A	I_L 92.389 A

5.

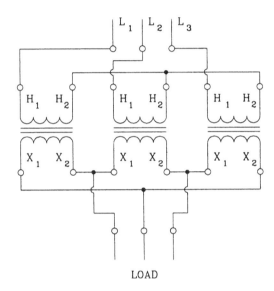

LOAD

6. 180 KVA

7. 84 KVA

DIRECT CURRENT MACHINES
Answers

1. C

2. A

3. D

4. B

5. A

6. D

7. C

8. B

9. B

10. A

11. D

12. C

13. A

14. B

15. C

16. A

17. B

18. D

19. B

20. C

21. A

22. C

23. D

24. B

25. A

THREE-PHASE MOTORS
Answers

1. A. Squirrel cage induction
   B. Wound rotor induction
   C. Synchronous

2. The speed of the rotating magnetic field

3. A. The number of stator poles per phase
   B. The frequency of the line

4. 6

5. The type of squirrel cage rotor

6. A. Strength of the magnetic field of the stator

   B. Strength of the magnetic field of the rotor

   C. Phase angle difference between stator and rotor flux

7. Wound rotor induction

8. Synchronous

9. The resistance connected in the rotor winding causes the rotor current to be more in phase with the stator current, reducing the phase angle between rotor and stator flux.

10. A squirrel cage winding used for starting

11. Overexcitation of the rotor

12. DC

13. After

14. Change any two stator leads

15. Consequent pole motor

SINGLE-PHASE MOTORS

Answers

1. A. Resistance-start induction-run
   B. Capacitor-start induction-run
   C. Capacitor-start capacitor-run

2. By reversing the start winding leads in relation to the run winding leads

3. The capacitor causes a 90° phase shift between start winding current and run winding current.

4. It disconnects the start winding when the motor reaches about 75% of full speed.

5. Capacitor-start capacitor-run

6. Highest starting torque of any single-phase motor

7. Moving the brushes 15° on either side of the neutral plane

8. A. Number of stator poles
   B. Frequency

9. Universal

10. Universal

11. In series

12. Reversing the armature leads or the field leads, but not both

13. No

14. To produce a rotating magnetic field

15. Generally considered not reversible

16. High impedance in the stator windings

17. 72 RPM

Delmar's Standard Textbook of Electricity 2E
Transparency Masters

Contents

Figure 1-12
Centrifugal force causes an object to pull away from its axis point.
Copyright 1999 Delmar Publishers

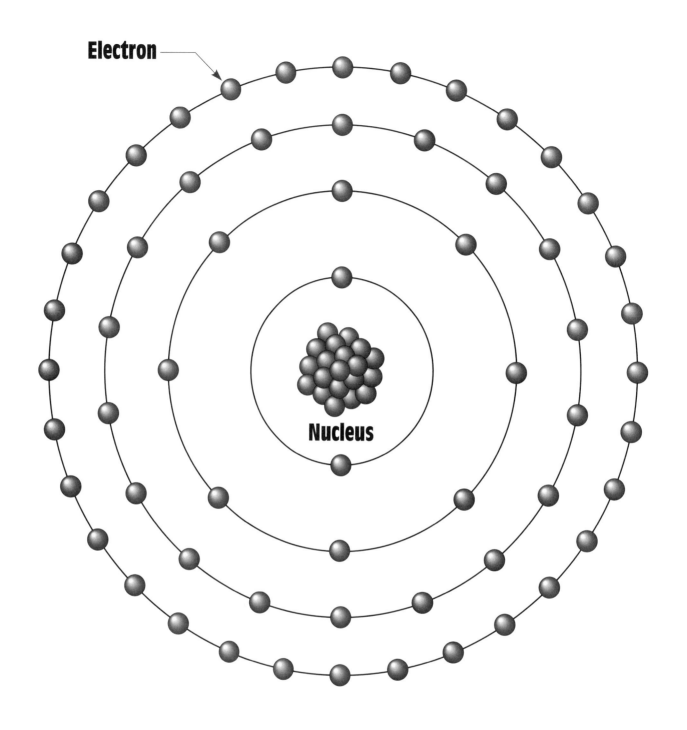

Figure 1-13

The first orbit or shell can contain a maximum of 2 electrons. The second orbit or shell can hold a maximum of 8 electrons, the third a maximum of 18 electrons, and the fourth a maximum of 32 electrons. No orbit or shell can contain more than 32 electrons. The formula 2N2 can be used to determine the maximum number of electrons in a particular orbit.

Copyright 1999 Delmar Publishers

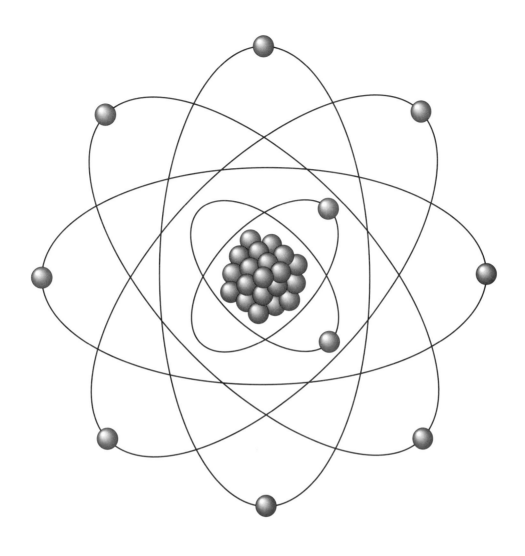

Figure 1-14
Electrons orbit the nucleus in a circular fashion.
Copyright 1999 Delmar Publishers

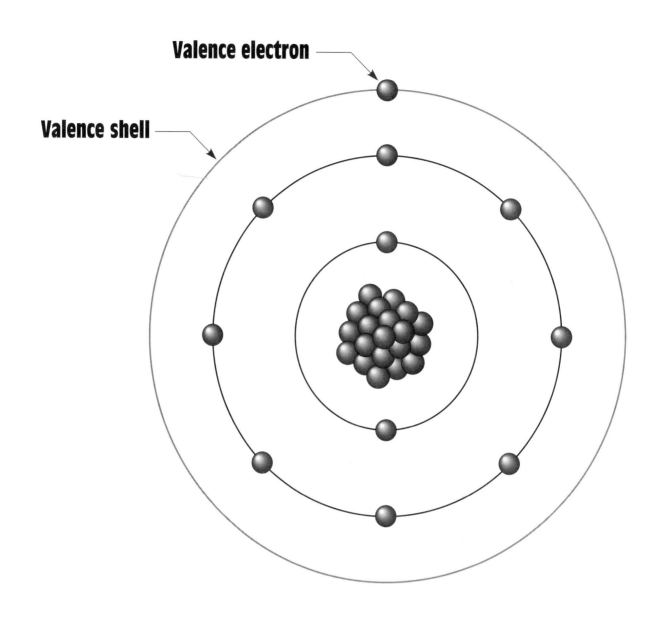

Valence electron

Valence shell

Figure 1-15
The electrons located in the outer orbit of an atom are valence electrons.
Copyright 1999 Delmar Publishers

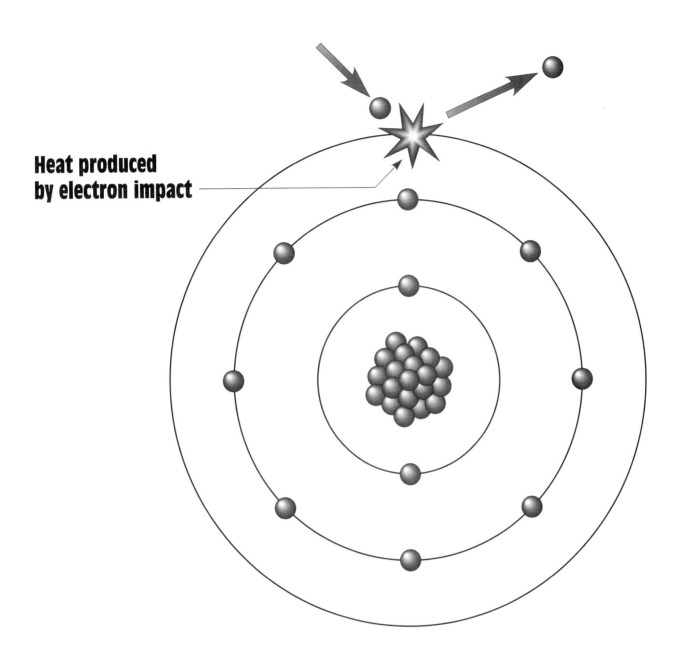

Heat produced by electron impact

Figure 1-17

The electron of one atom knocks another out of orbit.
Copyright 1999 Delmar Publishers

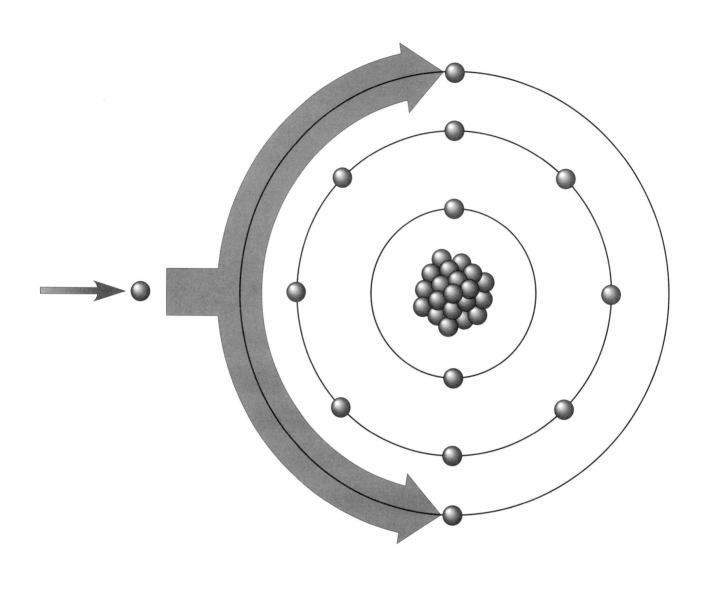

Figure 1-19
The energy of the striking electron is divided.
Copyright 1999 Delmar Publishers

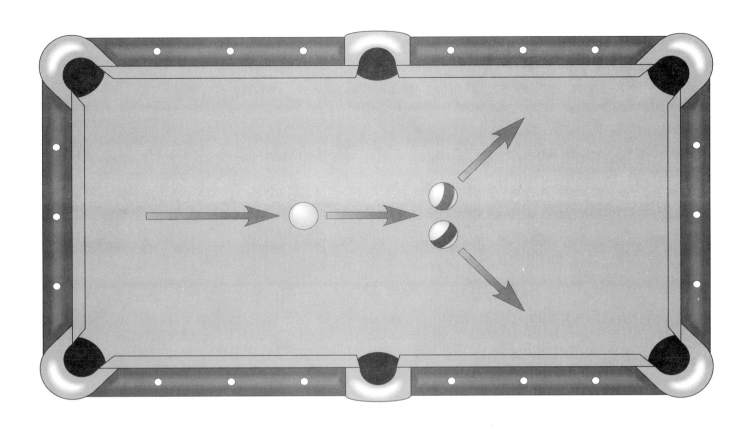

Figure 1-20
The energy of the cue ball is divided between the two other balls.
Copyright 1999 Delmar Publishers

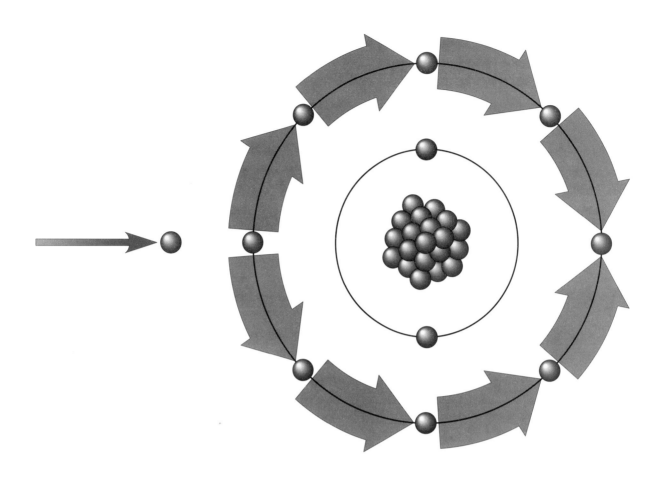

Figure 1-21

The energy of the striking electron is divided among the eight electrons.
Copyright 1999 Delmar Publishers

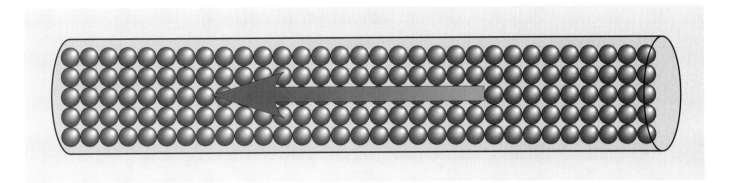

Figure 2-1

One amp equals one coulomb per second.

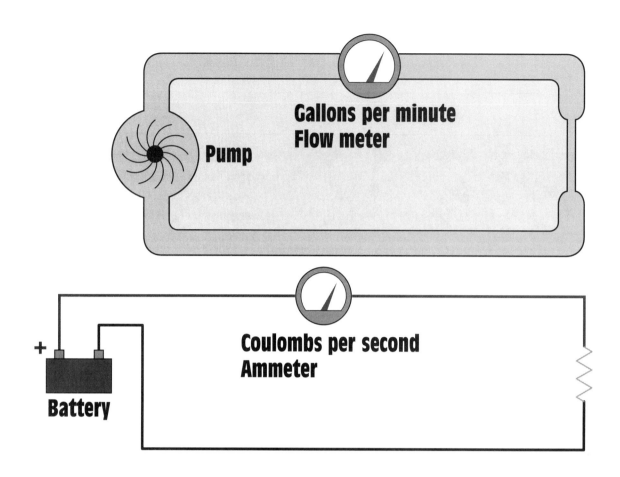

Figure 2-2

Current in an electrical circuit can be compared to flow rate in a water system.
Copyright 1999 Delmar Publishers

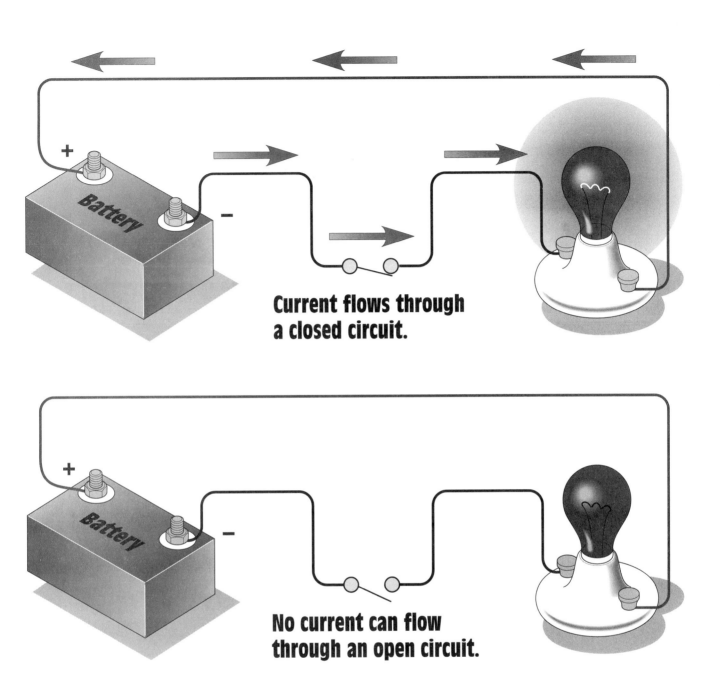

Current flows through a closed circuit.

No current can flow through an open circuit.

Figure 2-8

Current flows only through a closed circuit.
Copyright 1999 Delmar Publishers

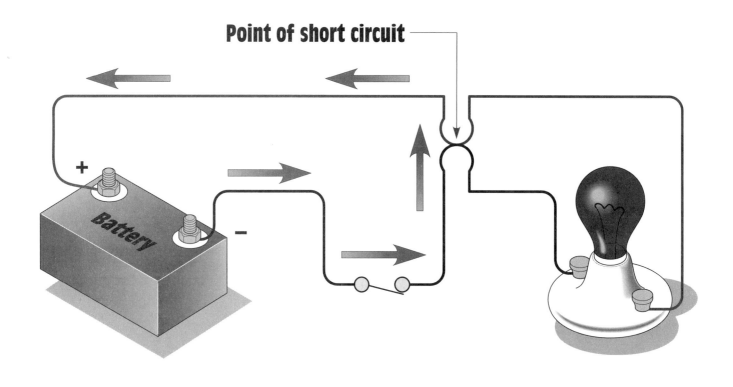

Point of short circuit

Figure 2-9

A short circuit bypasses the load and permits too much current to flow.
Copyright 1999 Delmar Publishers

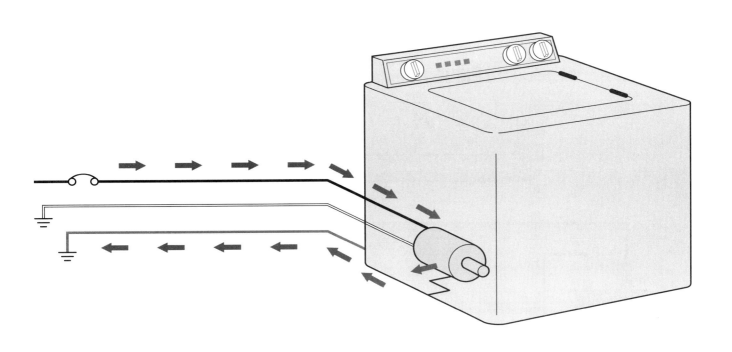

Figure 2-11
The grounding conductor provides a low-resistance path to ground.
Copyright 1999 Delmar Publishers

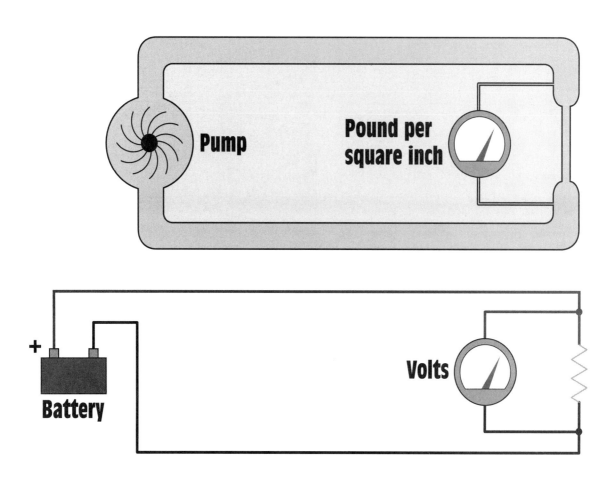

Figure 2-12
Voltage in an electrical circuit can be compared to pressure in a water system.
Copyright 1999 Delmar Publishers

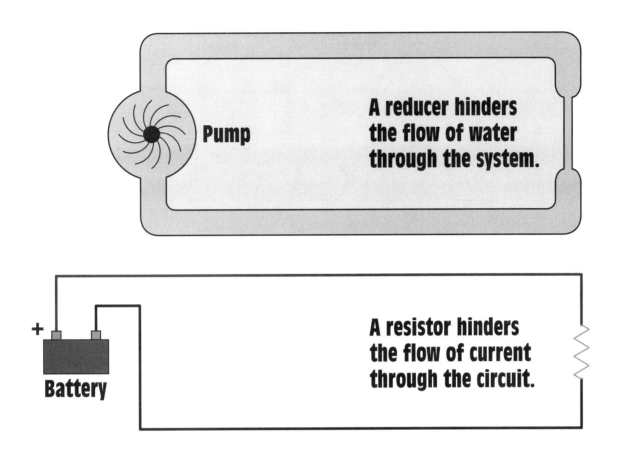

Figure 2-13
A resistor in an electrical circuit can be compared to a reducer in a water system.
Copyright 1999 Delmar Publishers

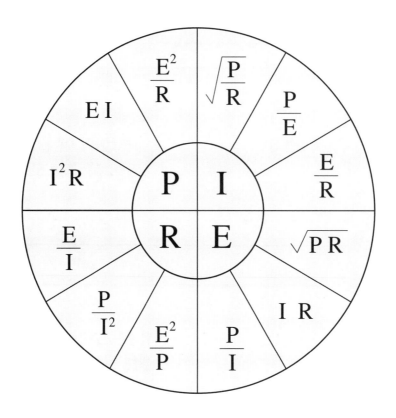

Figure 2-20

Formula chart for finding values of voltage, current, resistance, and power..
Copyright 1999 Delmar Publishers

Figure 4-1
The first compass was made by placing a small piece of magnetite on a piece of wood floating in a dish of water.
Copyright 1999 Delmar Publishers

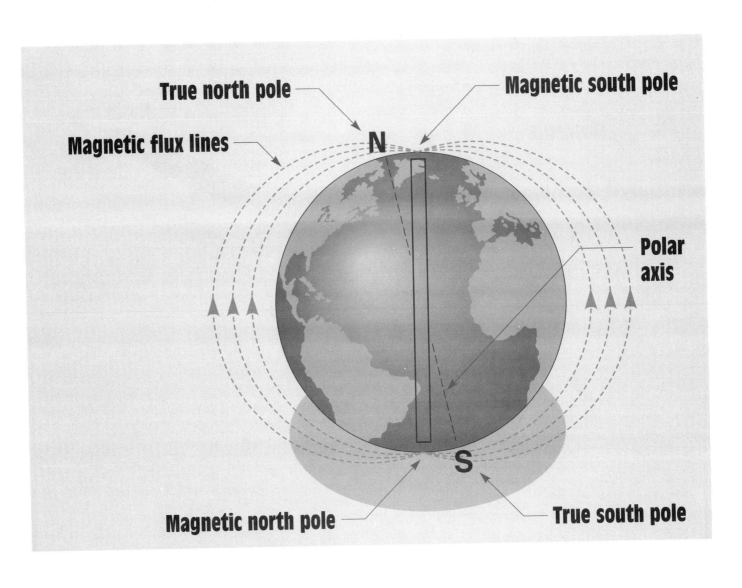

Figure 4-2

The earth is a magnet.
Copyright 1999 Delmar Publishers

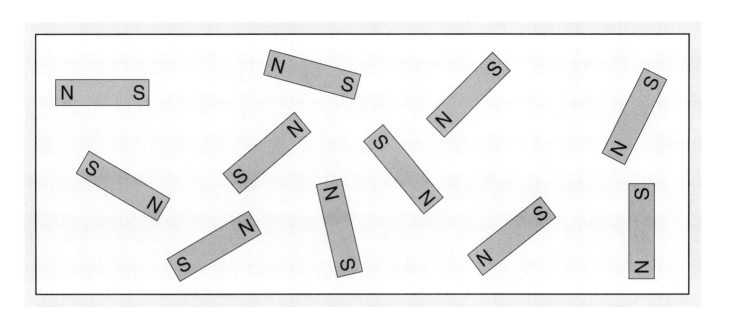

Figure 4-5
The atoms are disarrayed in a piece of nonmagnetized metal.
Copyright 1999 Delmar Publishers

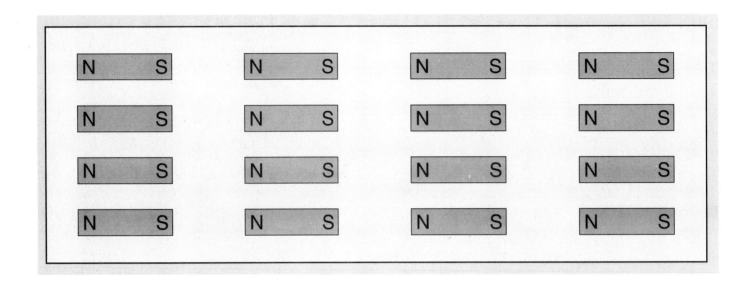

Figure 4-6
The atoms are aligned in an orderly fashion in a piece of magnetized metal.
Copyright 1999 Delmar Publishers

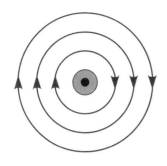

Current flowing out

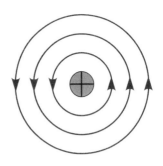

Current flowing in

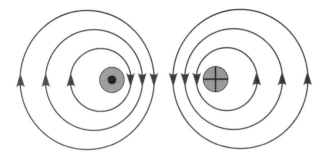

**Current flowing through
two conductors in the
opposite direction**

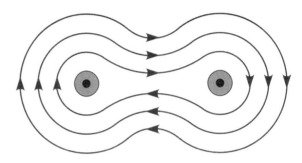

**Current flowing through
two conductors in the
same direction**

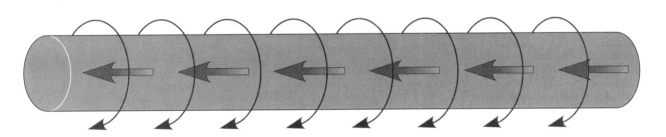

Figure 4-13
Current flowing through a conductor produces a magnetic field around the conductor.
Copyright 1999 Delmar Publishers

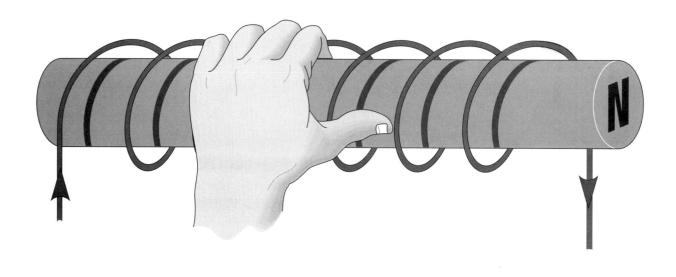

Figure 4-17
The left-hand rule can be used to determine the polarity of an electromagnet.
Copyright 1999 Delmar Publishers

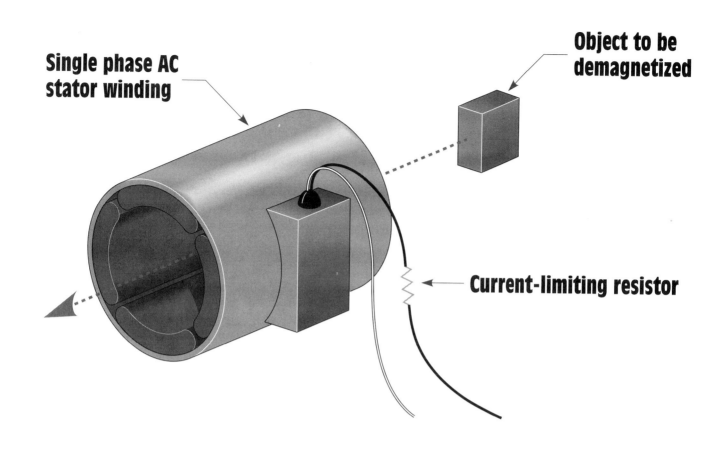

Single phase AC stator winding

Object to be demagnetized

Current-limiting resistor

Figure 4-18
Demagnetizing an object using a magnetic field produced by an alternating current.
Copyright 1999 Delmar Publishers

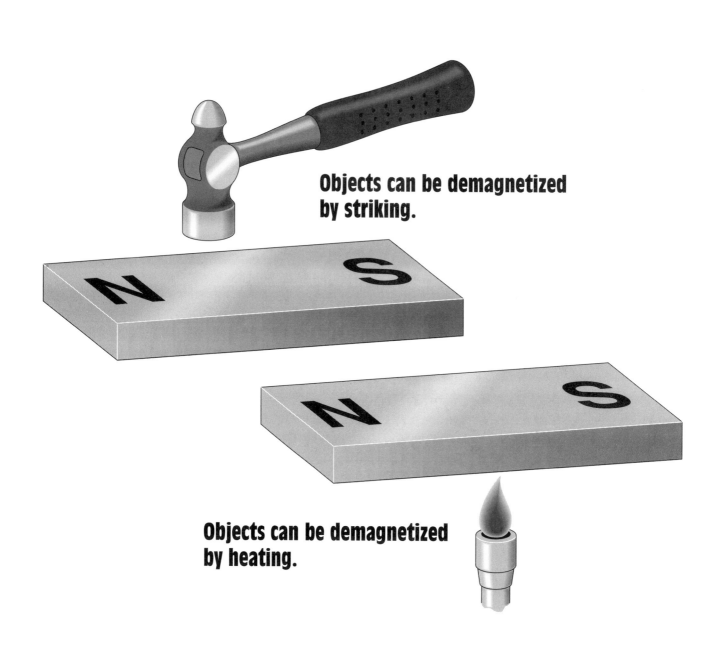

Objects can be demagnetized by striking.

Objects can be demagnetized by heating.

Figure 4-19
Other methods for demagnetizing objects.
Copyright 1999 Delmar Publishers

STANDARD RESISTANCE VALUES (Ω)

0.1% 0.25% 0.5%	1%	0.1% 0.25% 0.5%	1%	0.1% 0.25% 0.5%	1%	0.1% 0.25% 0.5%	1%	0.1% 0.25% 0.5%	1%
10.0	10.0	17.2	–	29.4	29.4	50.5	–	86.6	86.6
10.1	–	17.4	17.4	29.8	–	51.1	51.1	87.6	–
10.2	10.2	17.6	–	30.1	30.1	51.7	–	88.7	88.7
10.4	–	17.8	17.8	30.5	–	52.3	52.3	89.8	–
10.5	10.5	18.0	–	30.9	30.9	53.0	–	90.9	90.9
10.6	–	18.2	18.2	31.2	–	53.6	53.6	92.0	–
10.7	10.7	18.4	–	31.6	31.6	54.2	–	93.1	93.1
10.9	–	18.7	18.7	32.0	–	54.9	54.9	94.2	–
11.0	11.0	18.9	–	32.4	32.4	55.6	–	95.3	95.3
11.1	–	19.1	19.1	32.8	–	56.2	56.2	96.5	–
11.3	11.3	19.3	–	33.2	33.2	56.9	–	97.6	97.6
11.4	–	19.6	19.6	33.6	–	57.6	57.6	98.8	–
11.5	11.5	19.8	–	34.0	34.0	58.3	–		
11.7	–	20.0	20.0	34.4	–	59.0	59.0		
11.8	11.8	20.3	–	34.8	34.8	59.7	–		
12.0	–	20.5	20.5	35.2	–	60.4	60.4		
12.1	12.1	20.8	–	35.7	35.7	61.2	–		
12.3	–	21.0	21.0	36.1	–	61.9	61.9		
12.4	12.4	21.3	–	36.5	36.5	62.6	–		
12.6	–	21.5	21.5	37.0	–	63.4	63.4		
12.7	12.7	21.8	–	37.4	37.4	64.2	–	2%,5%	10%
12.9	–	22.1	22.1	37.9	–	64.9	64.9	10	10
13.0	13.0	22.3	–	38.3	38.3	65.7	–	11	–
13.2	–	22.6	22.6	38.8	–	66.5	66.5	12	12
13.3	13.3	22.9	–	39.2	39.2	67.3	–	13	–
13.5	–	23.2	23.2	39.7	–	68.1	68.1	15	15
13.7	13.7	23.4	–	40.2	40.2	69.0	–	16	–
13.8	–	23.7	23.7	40.7	–	69.8	69.8	18	18
14.0	14.0	24.0	–	41.2	41.2	70.6	–	20	–
14.2	–	24.3	24.3	41.7	–	71.5	71.5	22	22
14.3	14.3	24.6	–	42.2	42.2	72.3	–	24	–
14.5	–	24.9	24.9	42.7	–	73.2	73.2	27	27
14.7	14.7	25.2	–	43.2	43.2	74.1	–	30	–
14.9	–	25.5	25.5	43.7	–	75.0	75.0	33	33
15.0	15.0	25.8	–	44.2	44.2	75.9	–	36	–
15.2	–	26.1	26.1	44.8	–	76.8	76.8	39	39
15.4	15.4	26.4	–	45.3	45.3	77.7	–	43	–
15.6	–	26.7	26.7	45.9	–	78.7	78.7	47	47
15.8	15.8	27.1	–	46.4	46.4	79.6	–	51	–
16.0	–	27.4	27.4	47.0	–	80.6	80.6	56	56
16.2	16.2	27.7	–	47.5	47.5	81.6	–	62	–
16.4	–	28.0	28.0	48.1	–	82.5	82.5	68	68
16.5	16.5	28.4	–	48.7	48.7	83.5	–	75	–
16.7	–	28.7	28.7	49.3	–	84.5	84.5	82	82
16.9	16.9	29.1	–	49.9	49.9	85.6	–	91	–

Figure 5-12
Standard resistance values.
Copyright 1999 Delmar Publishers

159

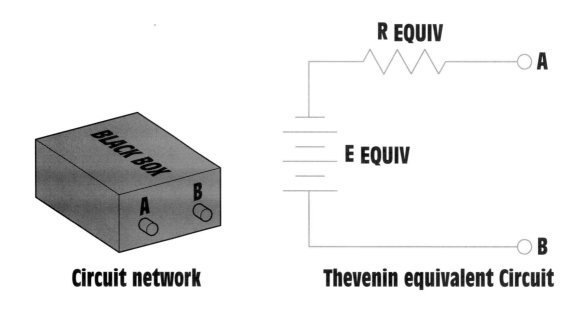

R EQUIV

A

E EQUIV

B

Circuit network   Thevenin equivalent Circuit

Figure 8-27
Thevenin's theorem reduces a circuit network to a single power source and a single series resistor.
Copyright 1999 Delmar Publishers

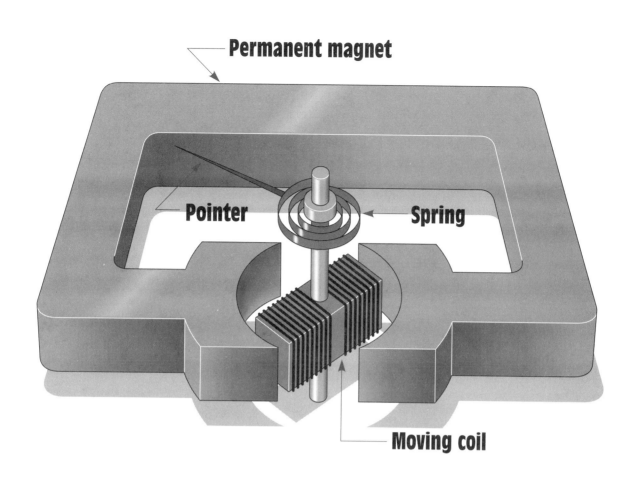

Figure 9-2
Basic d'Arsonval meter movement. This movement is often referred to as a moving coil meter movement.
Copyright 1999 Delmar Publishers

Permanent magnet

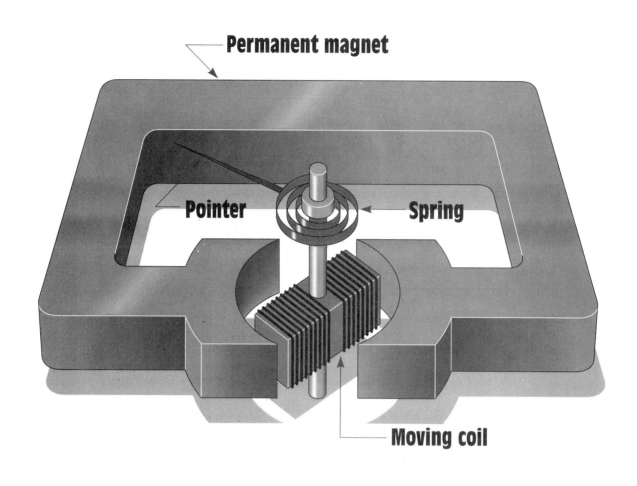

Pointer

Spring

Moving coil

Figure 9-2
Radial vane meter movement.
Copyright 1999 Delmar Publishers

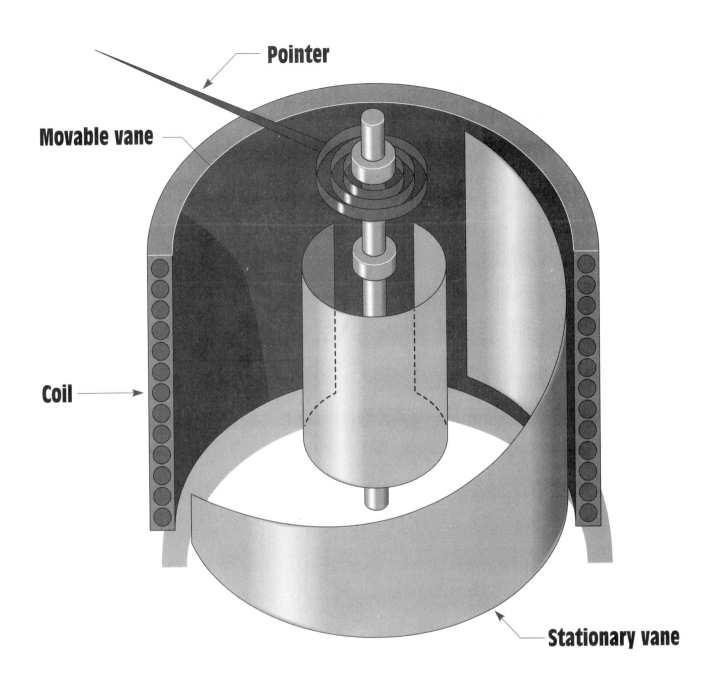

Pointer

Movable vane

Coil

Stationary vane

Figure 9-5
Concentric vane meter movement
Copyright 1999 Delmar Publishers

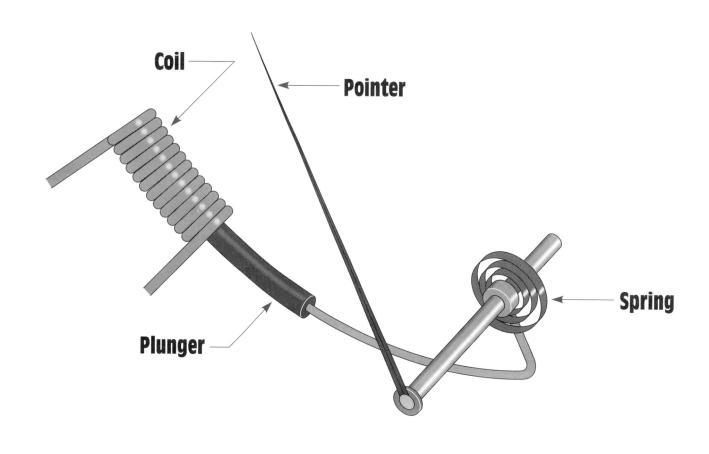

Figure 9-6
Plunger-type meter movement
Copyright 1999 Delmar Publishers

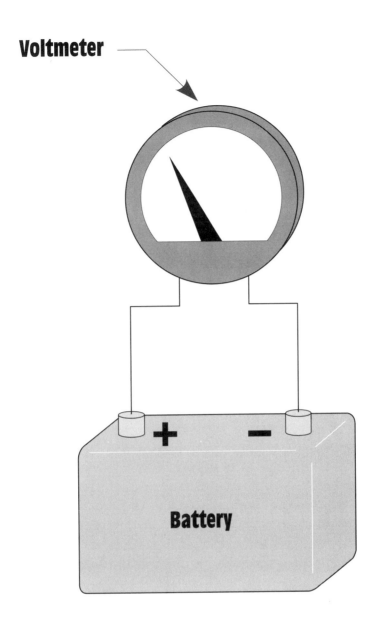

Voltmeter

Battery

Figure 9-7
A voltmeter connects directly across the power source.
Copyright 1999 Delmar Publishers

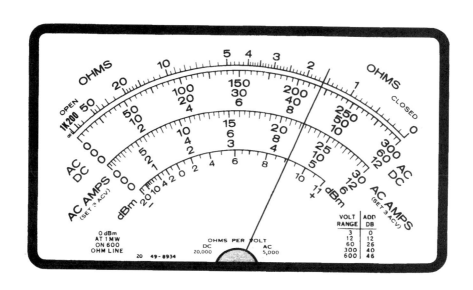

Figure 9-13
Scales of a typical multimeter.
Copyright 1999 Delmar Publishers

ELECTROMOTIVE SERIES OF METALS
(Partial list)

Lithium
Potassium
Sodium
Calcium
Magnesium
Aluminum
Manganese
Zinc
Chromium
Iron
Cadium
Cobalt
Nickel
Tin
Lead
Antimony
Copper
Mercury
Silver
Platinum
Gold

Figure 12-5
A partial list of the electromotive series of metals
Copyright 1999 Delmar Publishers

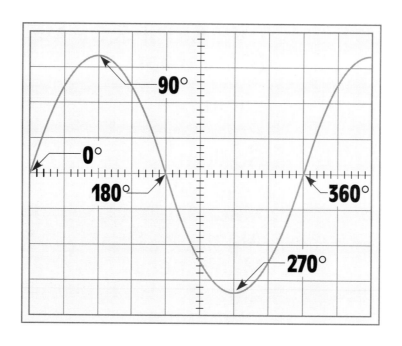

Figure 15-6
Sine wave
Copyright 1999 Delmar Publishers

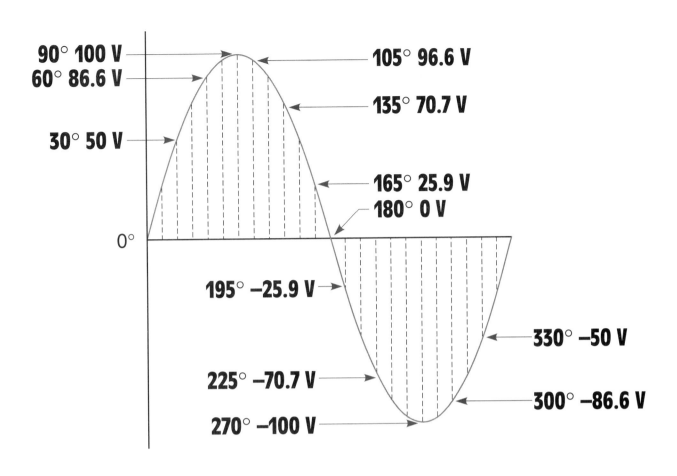

Figure 15-8
Instantaneous values of voltage along a sine wave
Copyright 1999 Delmar Publishers

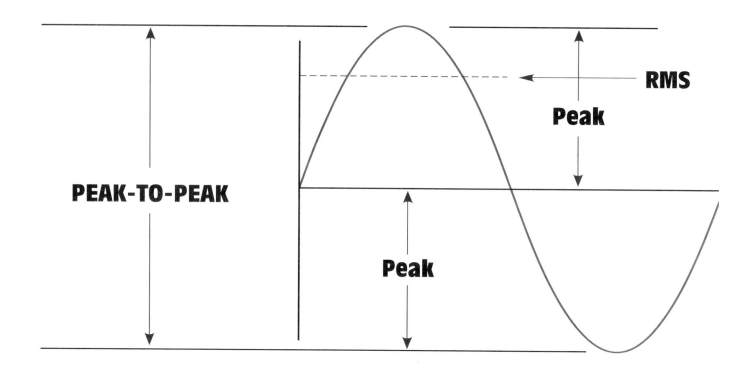

PEAK-TO-PEAK

RMS

Peak

Peak

Figure 15-9
Sine wave values
Copyright 1999 Delmar Publishers

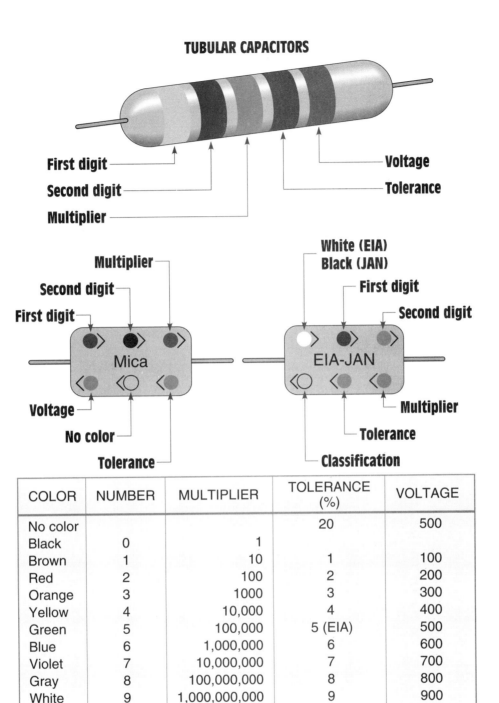

Figure 19-29
Identification of mica and tubular capacitors.
Copyright 1999 Delmar Publishers

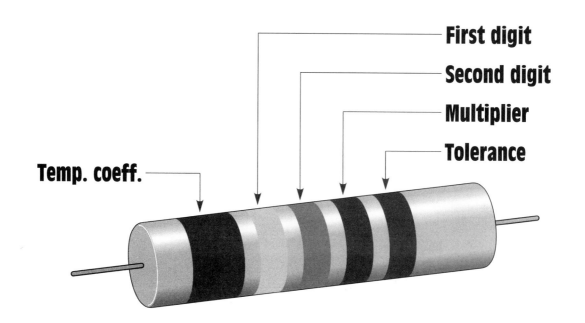

First digit

Second digit

Multiplier

Tolerance

Temp. coeff.

COLOR	NUMBER	MULTIPLIER	TOLERANCE OVER 10 pF	10 pF OR LESS	TEMP. COEFF.
Black	0	1	20%	2.0 pF	0
Brown	1	10	1%		N30
Red	2	100	2%		N80
Orange	3	1000			N150
Yellow	4				N220
Green	5				N330
Blue	6		5%	0.5 pF	N470
Violet	7				N750
Gray	8	0.01		0.25 pF	P30
White	9	0.1	10%	1.0 pF	P500

Figure 19-30
Color codes for ceramic capacitors
Copyright 1999 Delmar Publishers

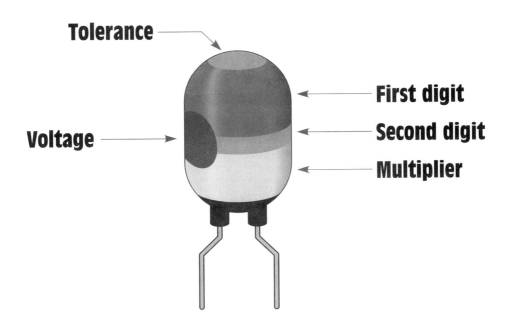

COLOR	NUMBER	MULTIPLIER	TOLERANCE (%)	VOLTAGE
			No dot 20	
Black	0			4
Brown	1			6
Red	2			10
Orange	3			15
Yellow	4	10,000		20
Green	5	100,000		25
Blue	6	1,000,000		35
Violet	7	10,000,000		50
Gray	8			
White	9			3
Gold			5	
Silver			10	

Figure 19-31
Dipped tantalum capacitors
Copyright 1999 Delmar Publishers

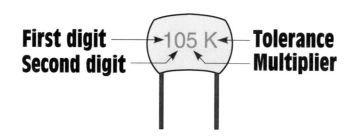

NUMBER	MULTIPLIER	TOLERANCE		
			10 pf or less	Over 10 pf
0	1	B	0.1 pf	
1	10	C	0.25 pf	
2	100	D	0.5 pf	
3	1000	F	1.0 pf	1%
4	10,000	G	2.0 pf	2%
5	100,000	H		3%
6		J		5%
7		K		10%
8	0.01	M		20%
9	0.1			

Figure 19-32
Film-type capacitors
Copyright 1999 Delmar Publishers

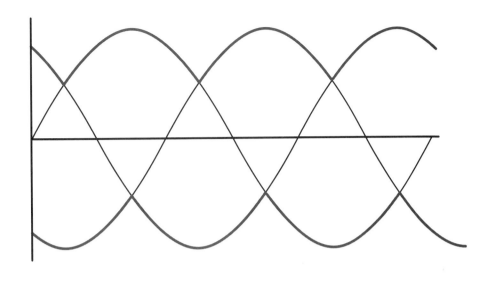

Figure 26-2
Three-phase power never falls to zero.
Copyright 1999 Delmar Publishers

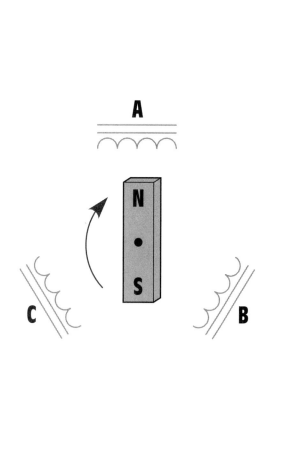

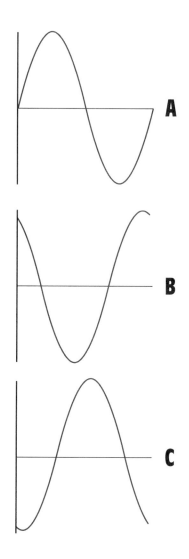

Figure 26-4
The voltages of a three-phase system are 120 degrees out of phase with each other.
Copyright 1999 Delmar Publishers

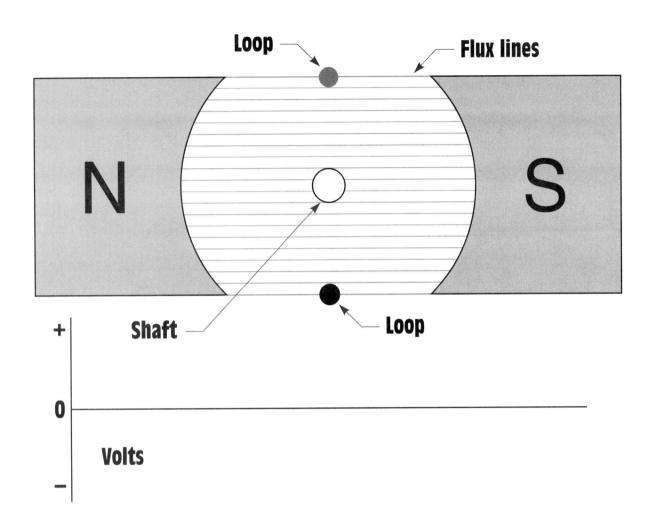

Figure 29-2
The loop is parallel to the lines of flux and no cutting action is taking place.
Copyright 1999 Delmar Publishers

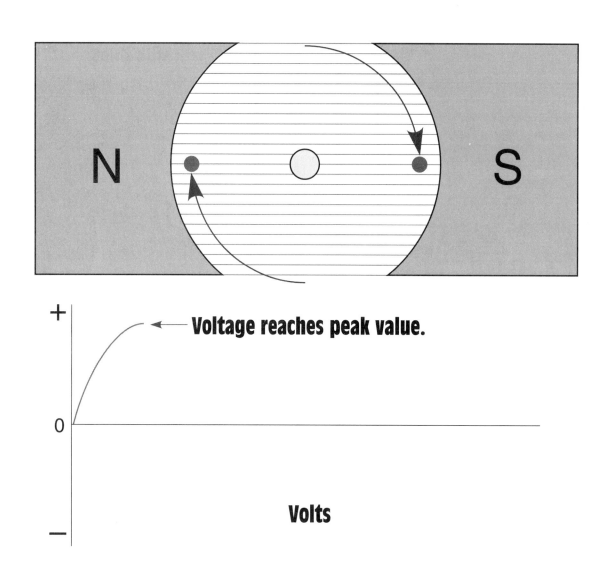

Figure 29-3
Induced voltage after 90 degrees of rotation
Copyright 1999 Delmar Publishers

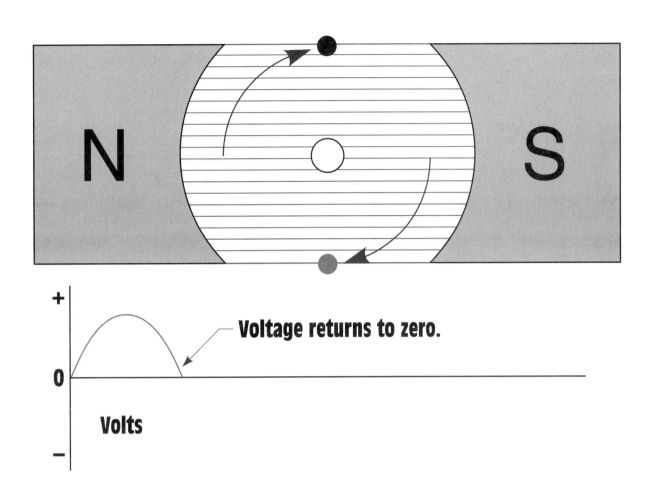

Figure 29-4
Induced voltage after 180 degrees of rotation
Copyright 1999 Delmar Publishers

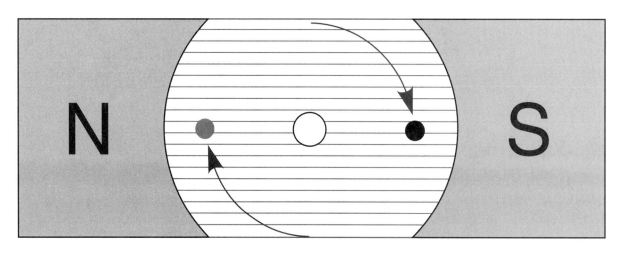

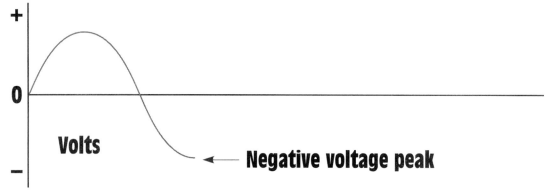

Figure 29-5
The negative voltage peak is reached after 270 degrees of rotation.

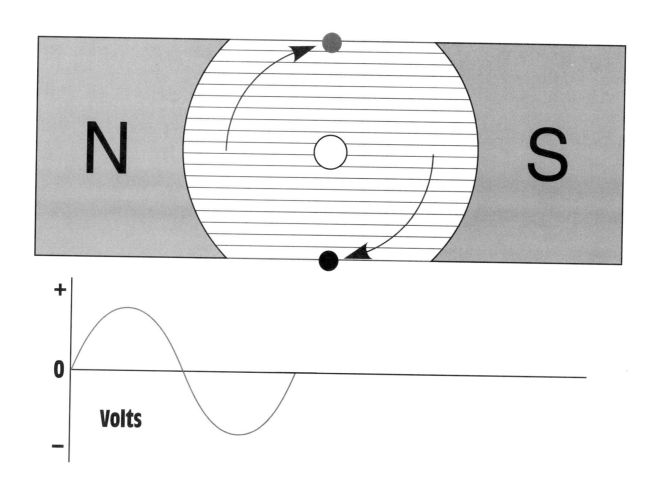

Figure 29-6
Voltage produced after 360 degrees of rotation.
Copyright 1999 Delmar Publishers

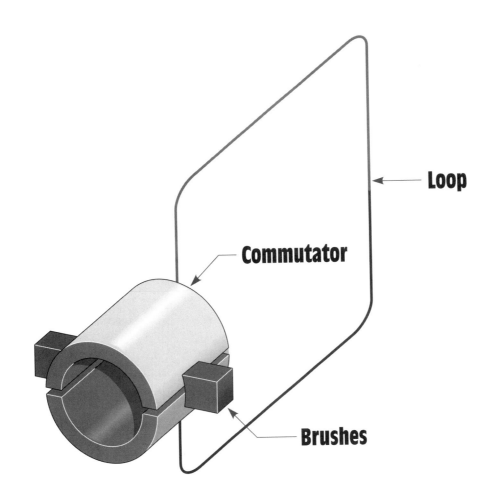

Figure 29-7

The commutator is used to convert the AC voltage produced in the armature into DC voltage.
Copyright 1999 Delmar Publishers

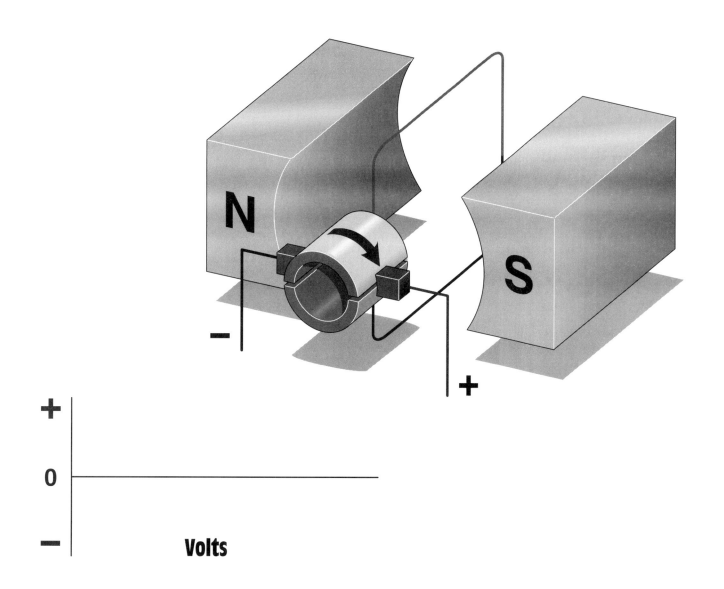

Figure 29-8
The loop is parallel to the lines of flux.
Copyright 1999 Delmar Publishers

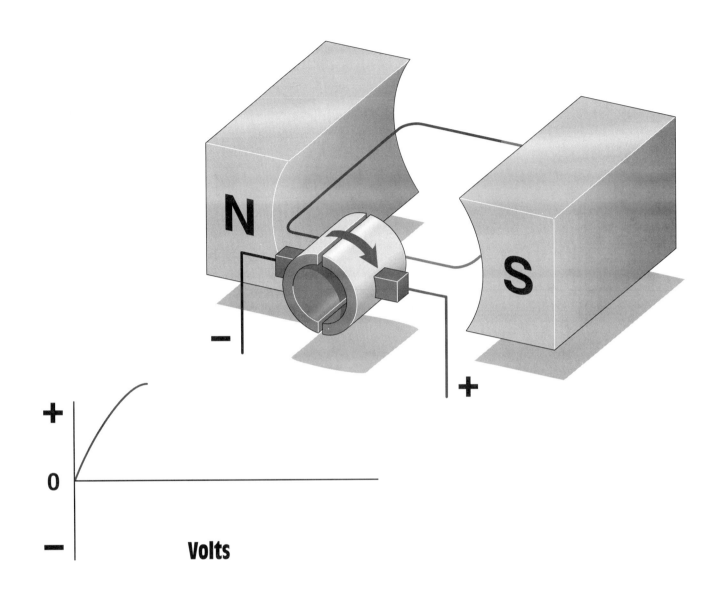

Figure 29-9
The loop has rotated 90 degrees.
Copyright 1999 Delmar Publishers

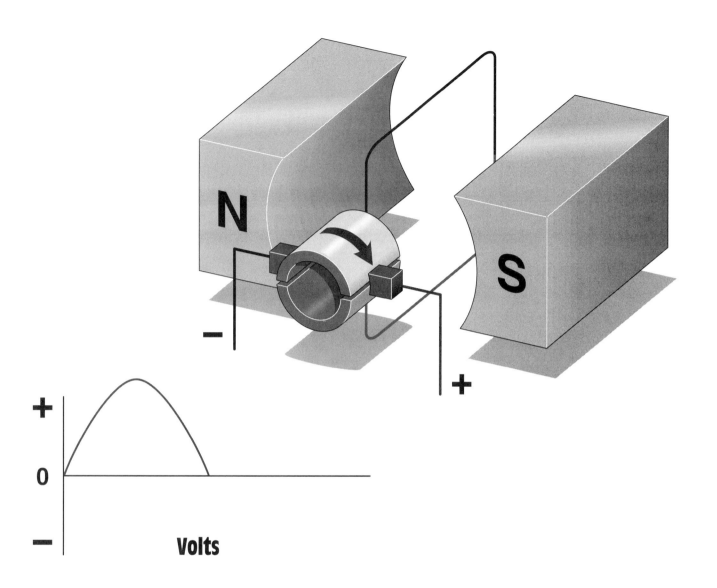

Figure 29-10
The loop has rotated 180 degrees.
Copyright 1999 Delmar Publishers

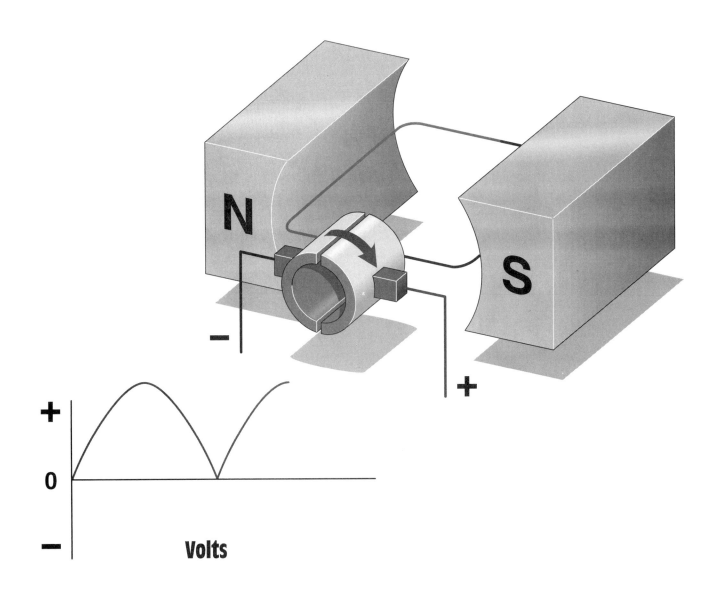

Figure 29-11
The commutator maintains the proper polarity.
Copyright 1999 Delmar Publishers

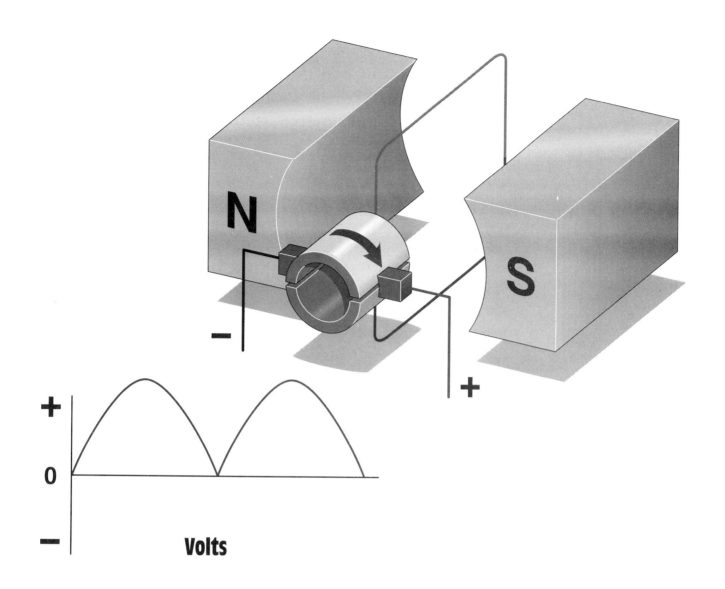

Figure 29-12
The loop completes one complete rotation.
Copyright 1999 Delmar Publishers

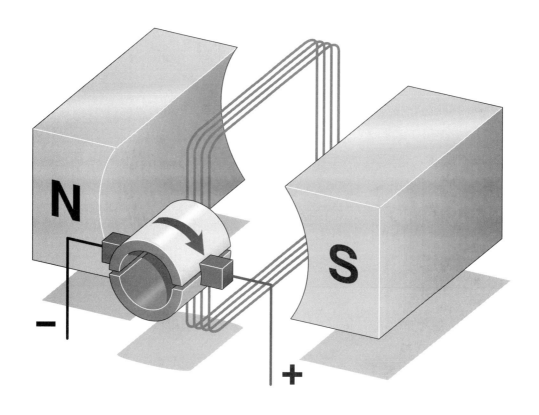

Figure 29-13
Increasing the number of turns increases the output voltage.
Copyright 1999 Delmar Publishers

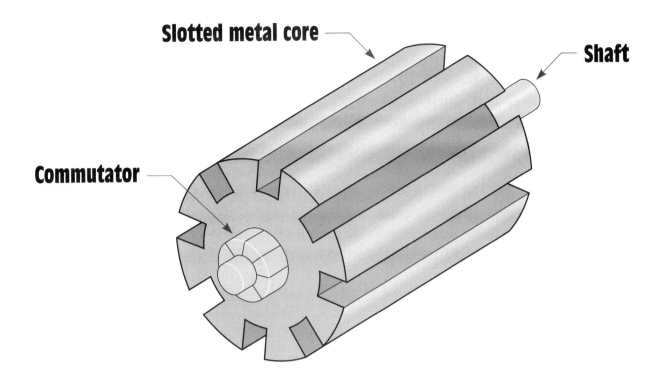

Figure 29-15
The loops of wire are wound around slots in a metal core.
Copyright 1999 Delmar Publishers

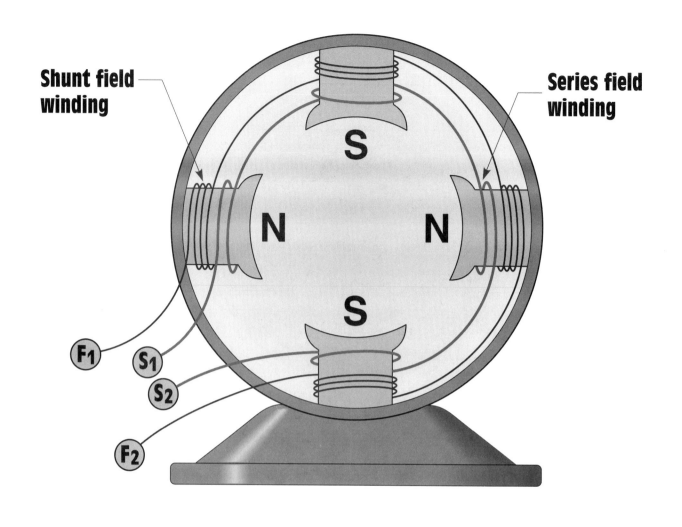

Figure 29-23
Both series and shunt field windings are contained on each pole piece.
Copyright 1999 Delmar Publishers

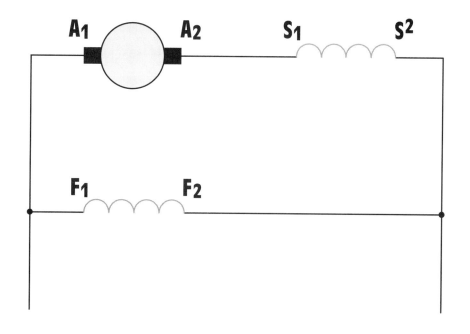

Figure 29-39
Schematic drawing of a long shunt compound generator
Copyright 1999 Delmar Publishers

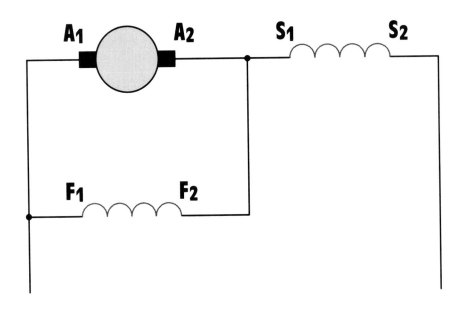

Figure 29-40
Schematic drawing of a short shunt compound generator
Copyright 1999 Delmar Publishers

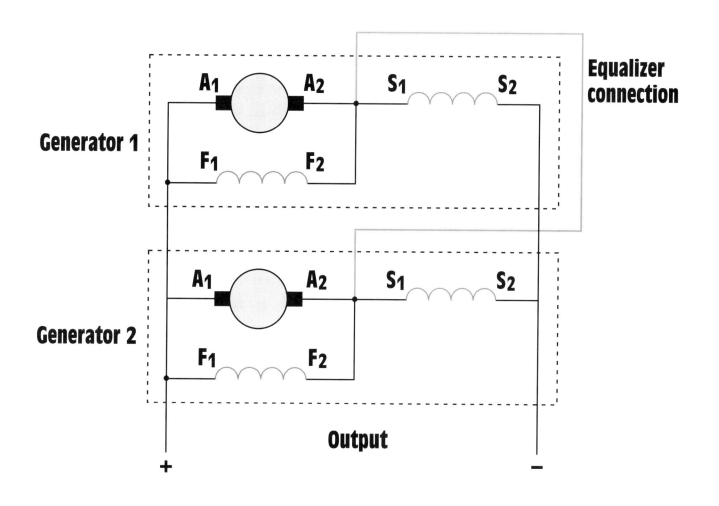

Figure 29-53
The equalizer connection is used to connect the series fields in parallel with each other.
Copyright 1999 Delmar Publishers

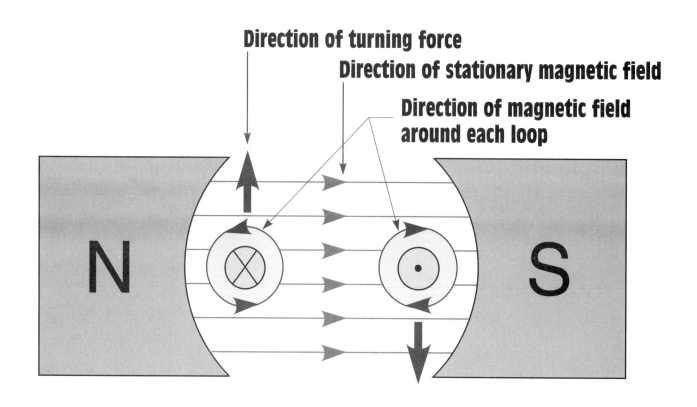

Direction of turning force

Direction of stationary magnetic field

Direction of magnetic field around each loop

Figure 30-2
Flux lines in the same direction repel each other, and flux lines
in the opposite direction attract each other.
Copyright 1999 Delmar Publishers

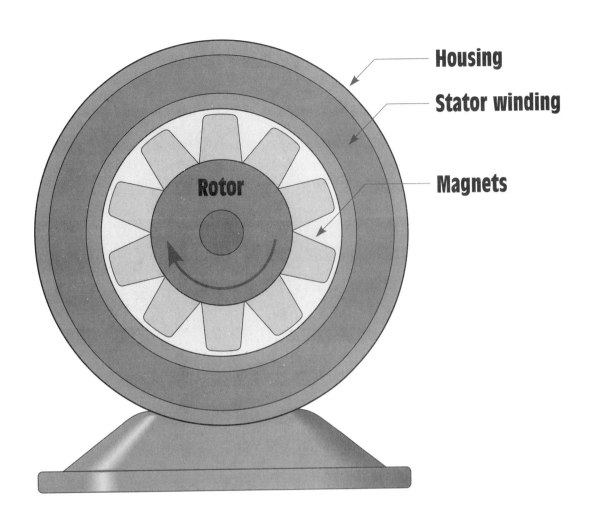

Housing

Stator winding

Magnets

Rotor

Figure 30-24
Brushless DC motor
Copyright 1999 Delmar Publishers

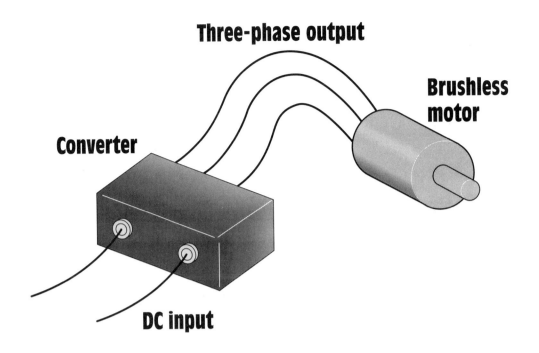

Three-phase output

Brushless motor

Converter

DC input

Figure 30-25
A converter changes direct current into three-phase alternating current.
Copyright 1999 Delmar Publishers

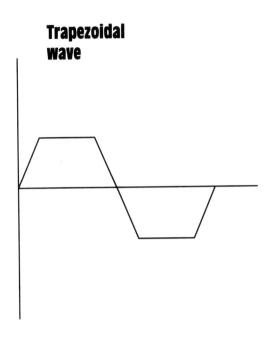

Trapezoidal wave

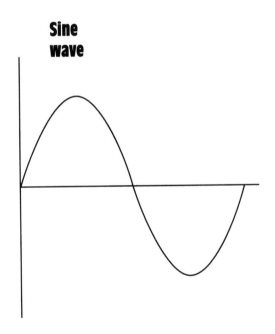

Sine wave

Figure 30-26
Most converters produce a trapezoidal wave, but some produce a sine wave.
Copyright 1999 Delmar Publishers

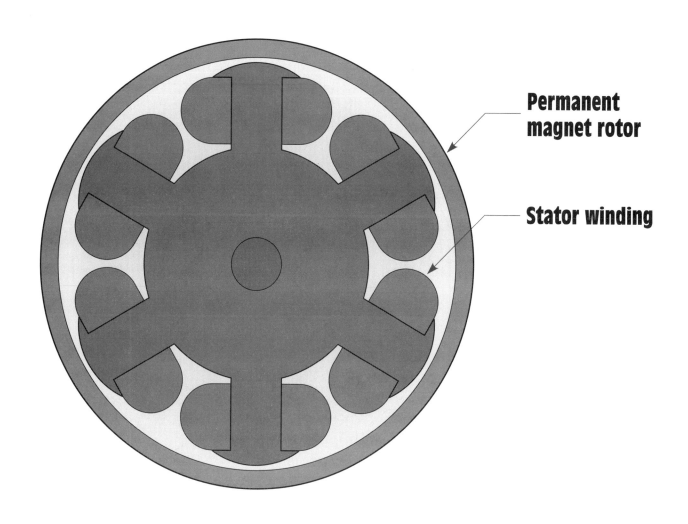

Permanent magnet rotor

Stator winding

Figure 30-27
Inside-out brushless DC motor
Copyright 1999 Delmar Publishers

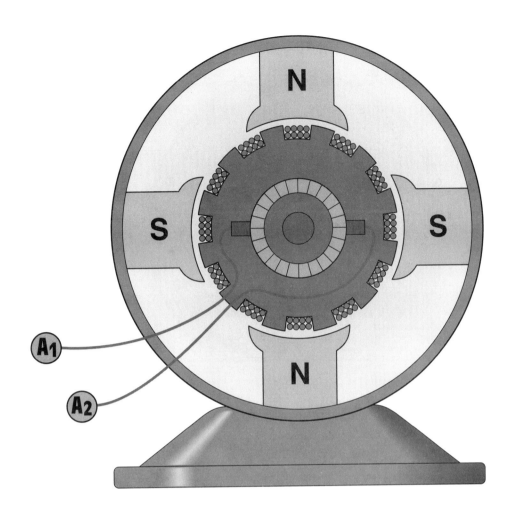

Figure 30-30
Permanent magnet motor
Copyright 1999 Delmar Publishers

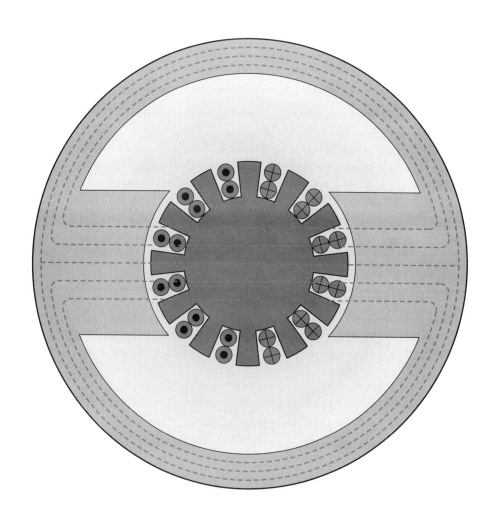

Figure 30-32
Magnetic lines of flux must travel a great distance between the poles of the permanent magnets.
Copyright 1999 Delmar Publishers

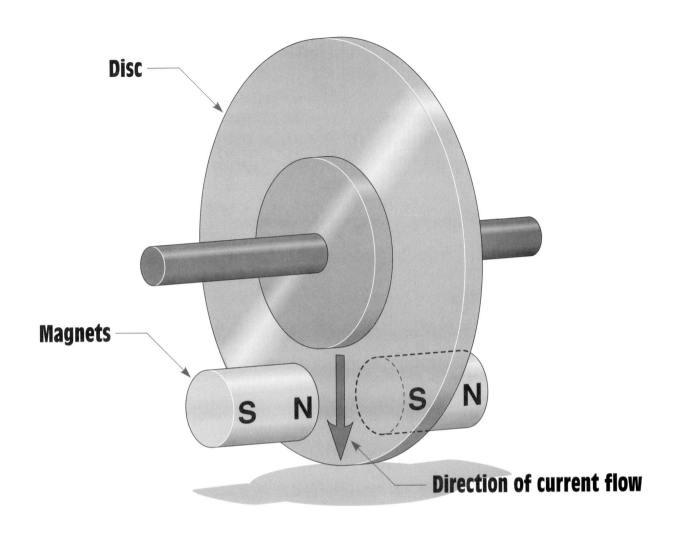

Figure 30-33
Basic construction of a ServoDisc motor
Copyright 1999 Delmar Publishers

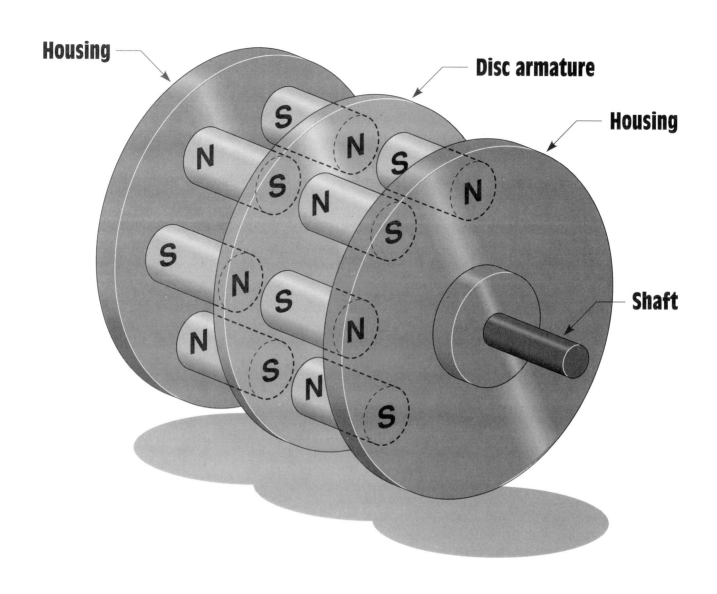

Figure 30-34
The permanent magnets are arranged to produce alternate magnetic polarities.
Copyright 1999 Delmar Publishers

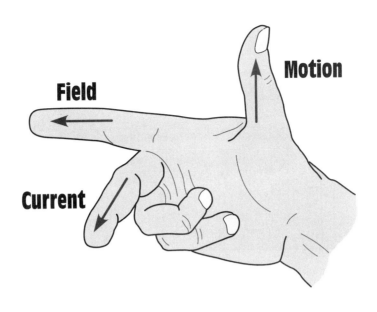

Figure 30-40
Right-hand motor rule
Copyright 1999 Delmar Publishers

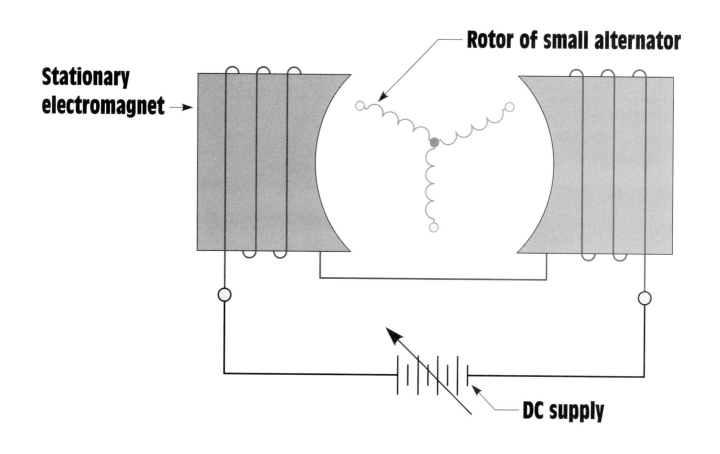

Stationary electromagnet →

Rotor of small alternator

DC supply

Figure 31-8
The brushless exciter uses stationary electromagnets.
Copyright 1999 Delmar Publishers

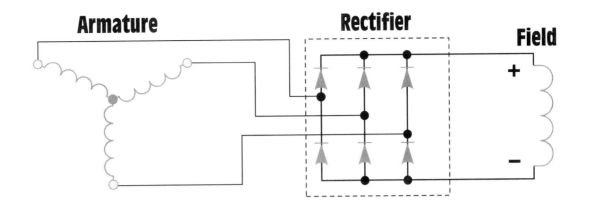

Figure 31-9
Basic brushless exciter circuit
Copyright 1999 Delmar Publishers

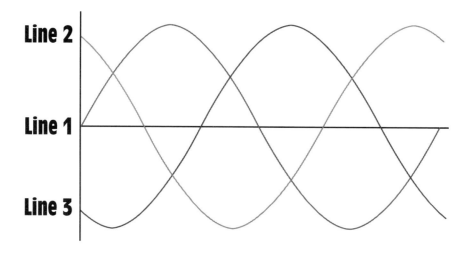

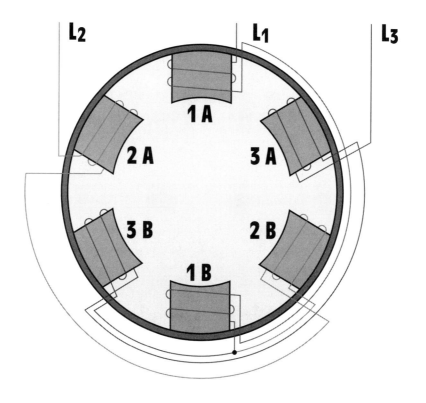

Figure 32-1
Three-phase stator and three sine wave voltages
Copyright 1999 Delmar Publishers

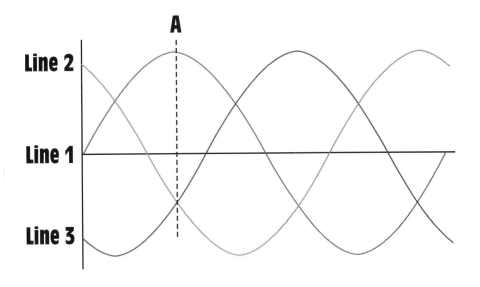

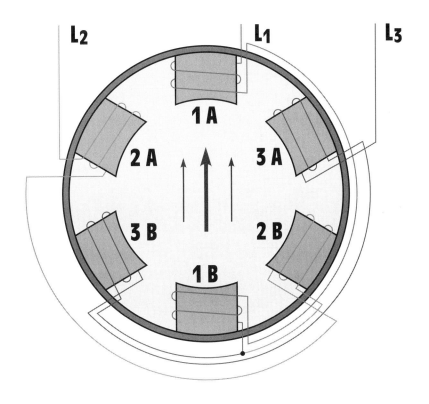

Figure 32-2
The magnetic field is concentrated between poles 1A and 1B.
Copyright 1999 Delmar Publishers

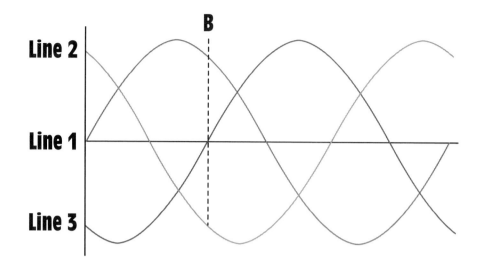

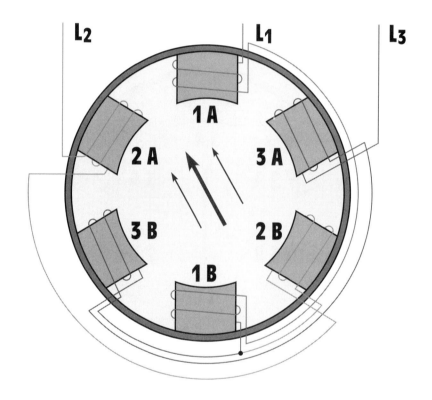

Figure 32-3
The magnetic field is concentrated between the poles of phases 1 and 2.
Copyright 1999 Delmar Publishers

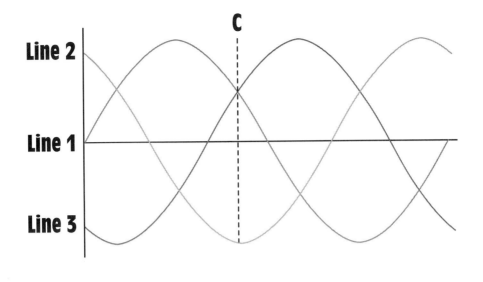

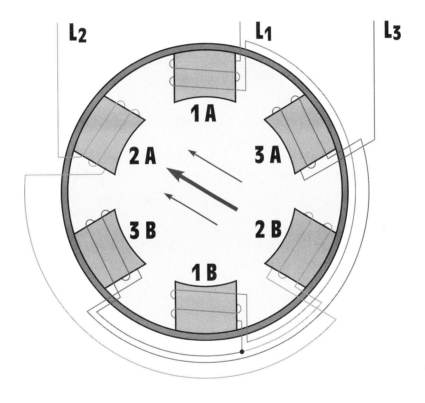

Figure 32-4
The magnetic field is concentrated between poles 2A and 2B.
Copyright 1999 Delmar Publishers

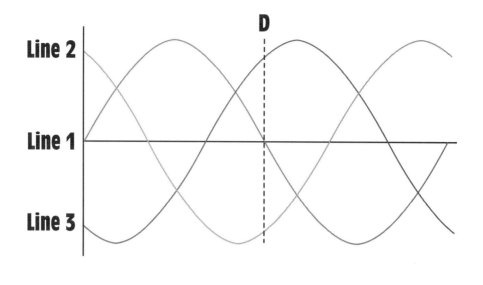

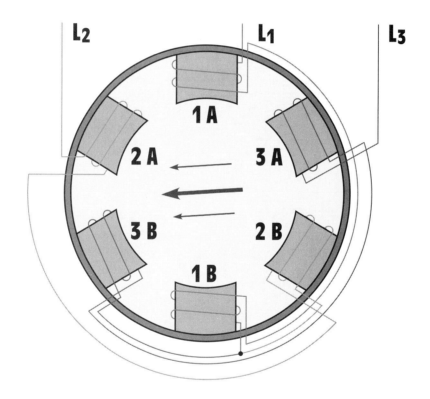

Figure 32-5
The magnetic field is concentrated between phases 2 and 3.
Copyright 1999 Delmar Publishers

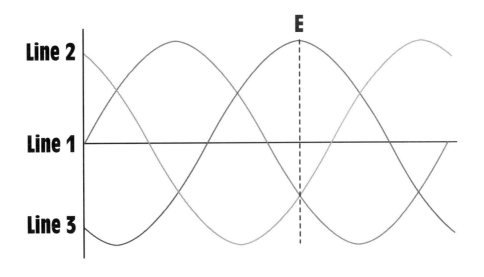

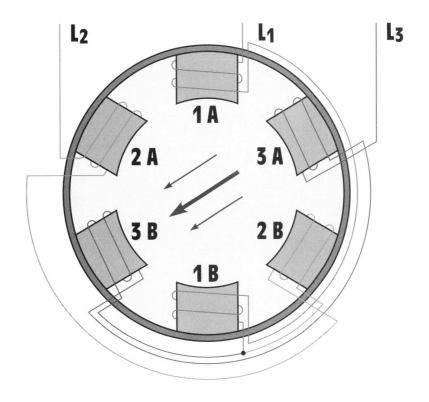

Figure 32-6
The magnetic field is concentrated between poles 3A and 3B.
Copyright 1999 Delmar Publishers

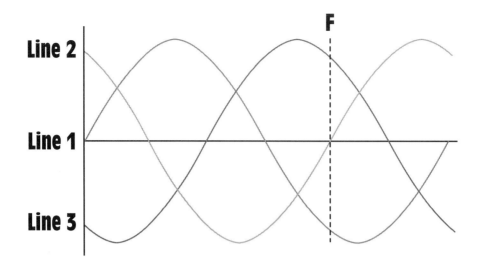

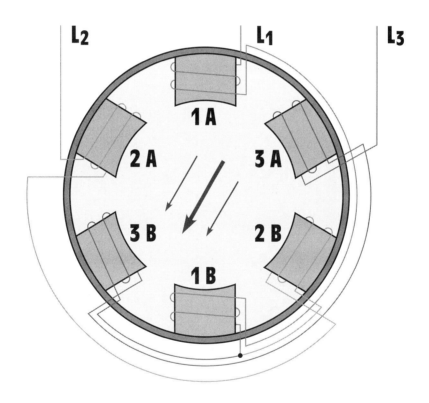

Figure 32-7
The magnetic field is concentrated between phases 1 and 3.
Copyright 1999 Delmar Publishers

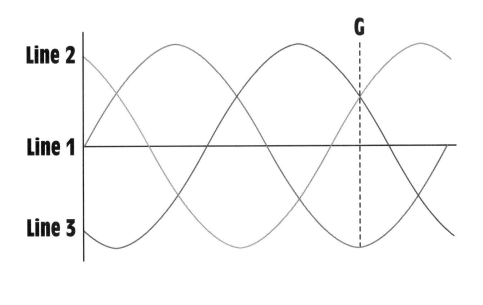

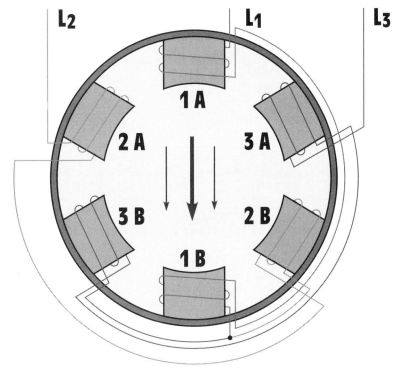

Figure 32-8
The magnetic field is concentrated between poles 1B and 1 A.
Copyright 1999 Delmar Publishers

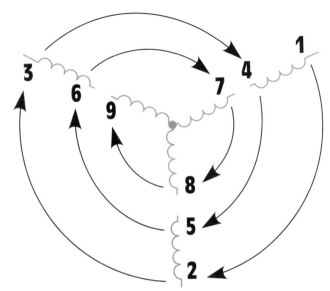

Standard numbering for a wye-connected motor

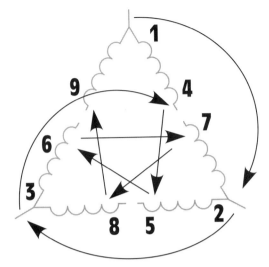

Standard numbering for a delta-connected motor

Figure 32-13
Standard numbering for three-phase motors
Copyright 1999 Delmar Publishers

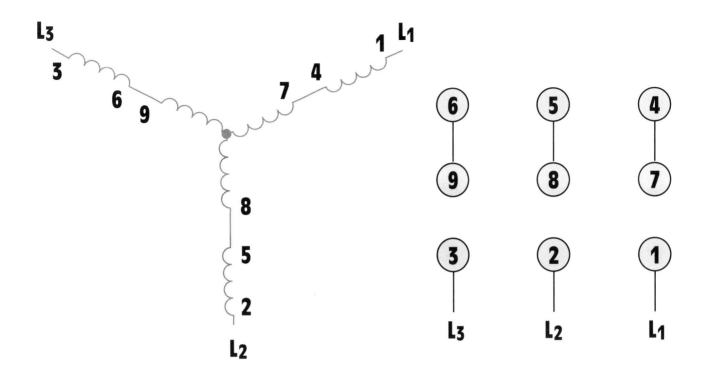

Figure 32-14
High-voltage wye connection
Copyright 1999 Delmar Publishers

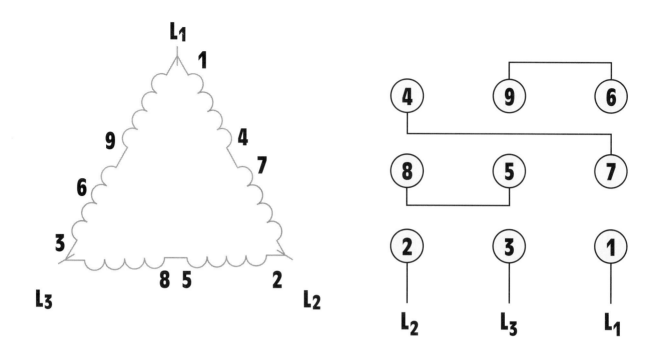

Figure 32-15

High-voltage delta connection

Copyright 1999 Delmar Publishers

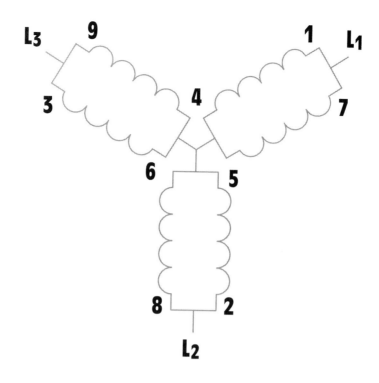

Figure 32-16
Stator windings connected in parallel
Copyright 1999 Delmar Publishers

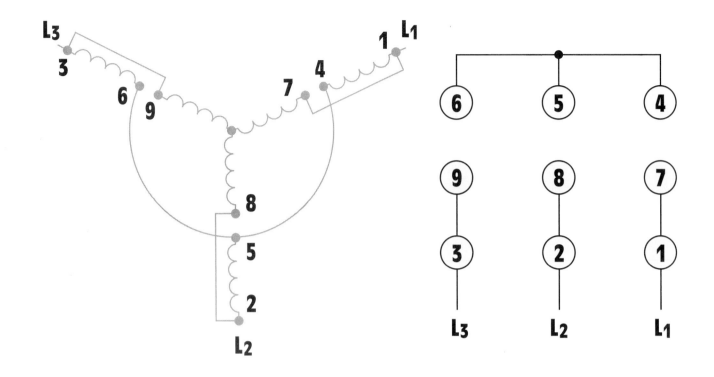

Figure 32-17
Low-voltage wye connection
Copyright 1999 Delmar Publishers

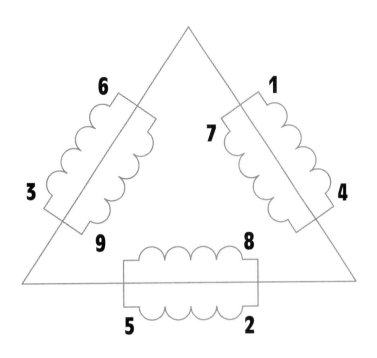

Figure 32-18
Parallel delta connection
Copyright 1999 Delmar Publishers

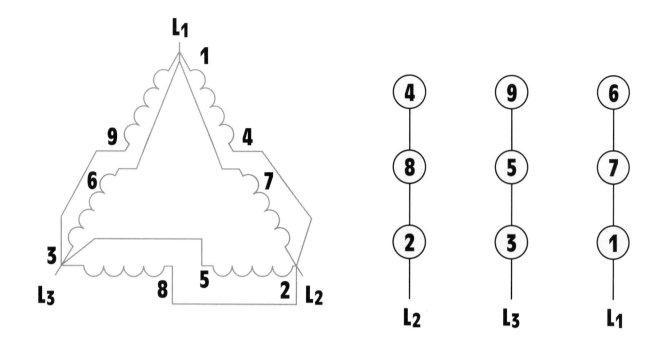

Figure 32-19
Low-voltage delta connection
Copyright 1999 Delmar Publishers

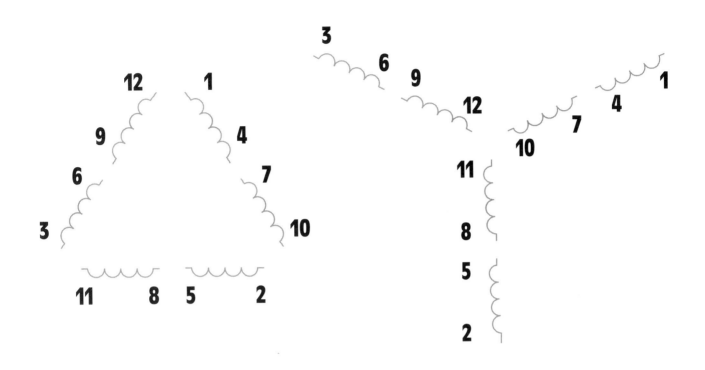

Figure 32-20
12-lead motor
Copyright 1999 Delmar Publishers

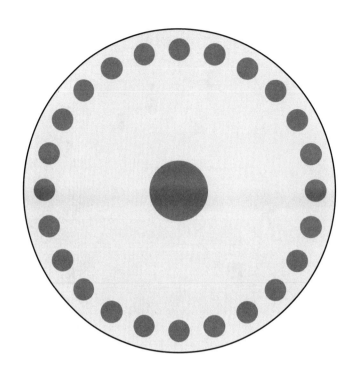

Figure 32-32
Type A rotor
Copyright 1999 Delmar Publishers

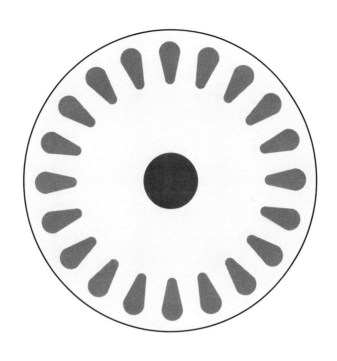

Figure 32-33
Type B-E rotor
Copyright 1999 Delmar Publishers

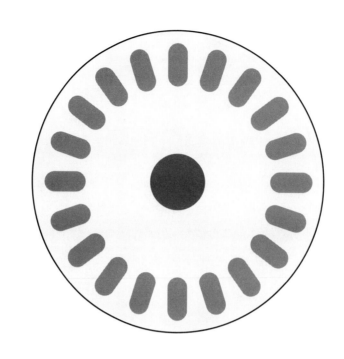

Figure 32-34
Type F-V rotor
Copyright 1999 Delmar Publishers

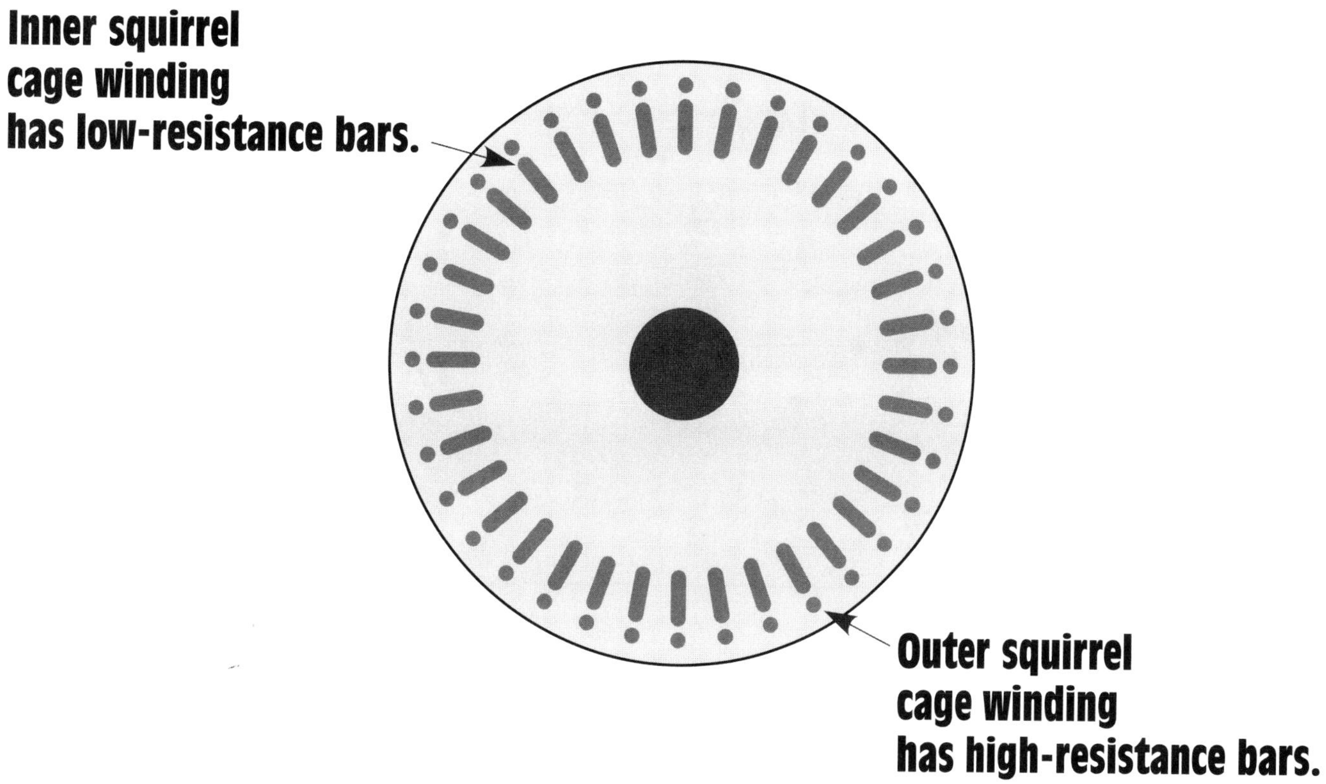

Inner squirrel cage winding has low-resistance bars.

Outer squirrel cage winding has high-resistance bars.

Figure 32-35
Double squirrel cage rotor
Copyright 1999 Delmar Publishers

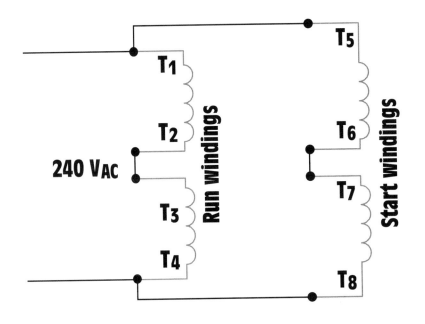

Figure 33-20
High-voltage connection for a split-phase motor with two run and two start windings
Copyright 1999 Delmar Publishers

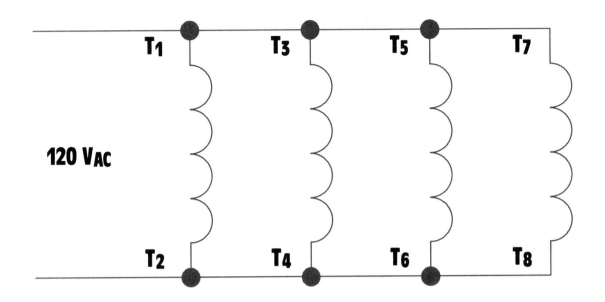

Figure 33-21
Low-voltage connection for a split-phase motor with two run and two start windings
Copyright 1999 Delmar Publishers

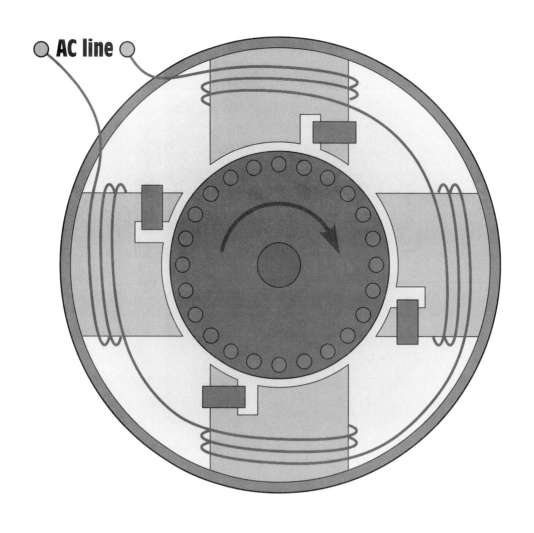

AC line

Figure 33-35
Four-pole shaded-pole induction motor
Copyright 1999 Delmar Publishers

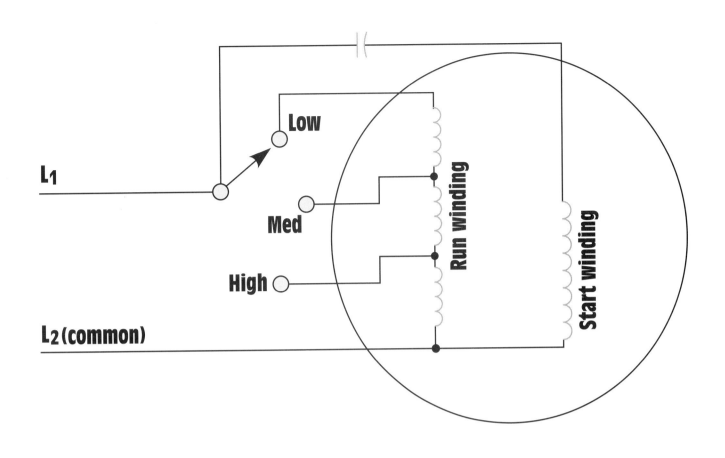

Figure 33-37
Three-speed motor
Copyright 1999 Delmar Publishers

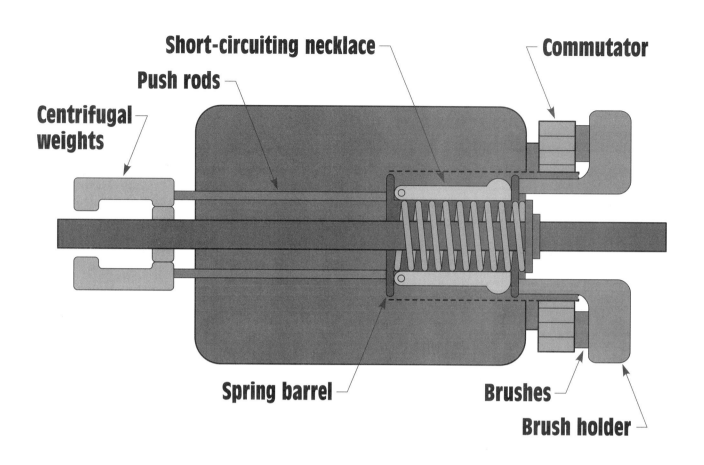

Centrifugal weights

Push rods

Short-circuiting necklace

Commutator

Spring barrel

Brushes

Brush holder

Figure 33-45
Brush-lifting type, repulsion-start induction-run motor
Copyright 1999 Delmar Publishers

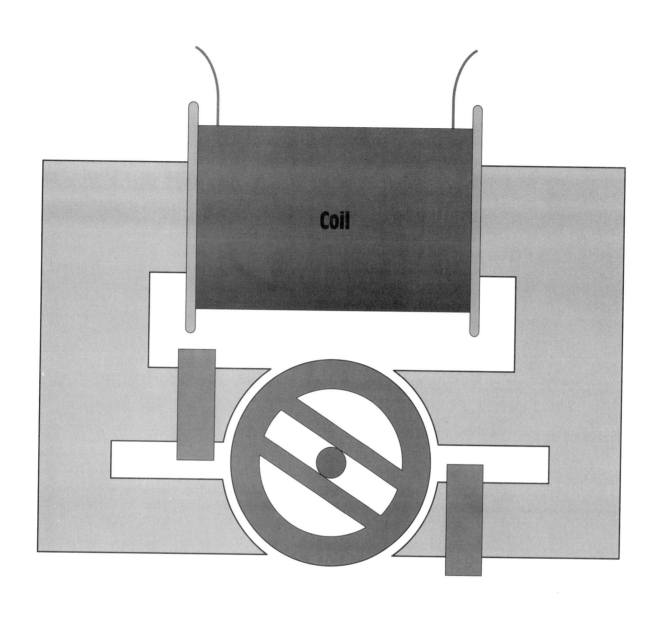

Figure 33-48
Warren motor
Copyright 1999 Delmar Publishers

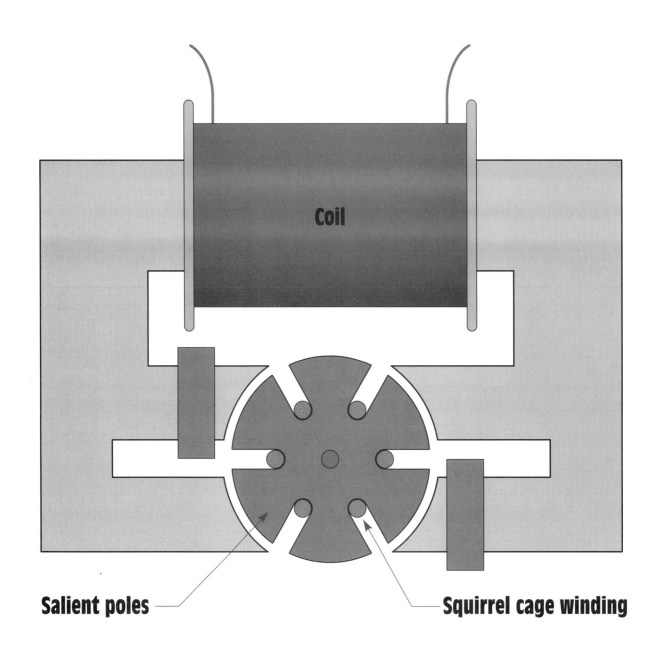

Coil

Salient poles

Squirrel cage winding

Figure 33-49
Holtz motor
Copyright 1999 Delmar Publishers